Oxide Thin Film Transistors

Oxide Thin Film Transistors

Yue Kuo
National Yang Ming Chiao Tung University

Hideo Hosono
Tokyo Institute of Technology

Michael S. Shur
Rensselaer Polytechnic Institute

Jin Jang
Kyung Hee University

WILEY

Published by John Wiley & Sons, Inc., Hoboken, New Jersey.
Published simultaneously in Canada.

Library of Congress Cataloging-in-Publication Data applied for:

Hardback ISBN: 9781394200146

Cover Design: Wiley
Cover Image: © Gile68/Getty Images

Set in 9.5/12.5pt STIXTwoText by Straive, Chennai, India

Contents

Preface

The thin film transistor (TFT)-based active-matrix flat panel display (AMFPD) industry is the second largest electronic industry only behind integrated circuit (IC) industry. The success of this industry is supported by the TFT technology that can be manufactured over large-area substrates at low temperatures with high throughputs.

Amorphous silicon (a-Si:H) TFTs have successfully dominated the large-area LCD product market since the mid-1980s. However, their low field effect mobility greatly limited applications requiring high-speed witching or large driving currents. Polycrystalline silicon (poly-Si) TFTs have mobilities more than an order of magnitude higher than those of a-Si:H TFTs. However, due to the complicated fabrication process, poly-Si TFTs are mainly used in small-sized displays and circuits.

Metal oxide TFTs, abbreviated as oxide TFTs, have emerged as the technology that can replace a-Si:H TFTs in large-area display applications and poly-Si TFTs in high-speed or high-current products. Oxide TFT arrays and various complicated circuits can be fabricated improving the performance, reliability, and production cost beyond AM LCDs.

The purpose of this book is to present a global view of the oxide TFT technology including fundamental material properties, device operation principles, fabrication processes, and applications. This book can be used as a textbook for students who would like to learn about oxide TFTs and as a reference for researchers and engineers who work on oxide TFTs and in related areas.

Yue Kuo acknowledges the award of Yushan Scholar sponsored by the Ministry of Education, Taiwan, ROC during the preparation of this book.

27 April 2024

Yue Kuo
Hideo Hosono
Michael S. Shur
Jin Jang

1

Introduction of Metal Oxide Thin Film Transistors

1.1 Progress of Thin Film Transistor Development

The concept of the field effect transistor (FET) was originally proposed in 1925 and patented in 1926, as shown in Figure 1.1 [1, 2]. The thin film transistor (TFT) is operated on the same principle as that of the metal oxide field effect transistor (MOSFET). The first functional TFT, which had an evaporated microcrystalline (mc) CdS semiconductor channel layer and SiO gate dielectric with a staggered structure, was reported by Weimer in 1962 [3]. It was originally designed for electronic circuit applications [4]. Subsequently, various types of inorganic or organic semiconductor materials including tin oxide were made into TFTs [5, 6]. Originally, TFTs were aimed at circuit applications. However, they could not compete with the silicon wafer-based MOSFETs. New applications of TFTs were searched based on the advantageous low-temperature, low-temperature process capability. Brody et al. [7] first reported a 6″ by 6″ active matrix liquid crystal display (AMLCD) with a CdSe TFT array. It opened the high-performance flat panel era because the TFT could provide gray-scale switching capability for LC in each pixel and a TFT array could be prepared at a low temperature on a large-area substrate.

In the beginning, there were many efforts in mass production of AMLCDs using TFT backplates, but only limited success was achieved due to many manufacturing issues. However, the report of the amorphous silicon (a-Si:H) TFT by LeComber et al. [8] in 1979 provided a viable solution in manufacturing the large-area AMLCDs. It immediately attracted great attention from all electronics companies. The intrinsic a-Si:H channel, n^+ ohmic contact layer, and gate dielectric in the TFT were all deposited by the plasma-enhanced chemical vapor deposition (PECVD) process. The low temperature process condition makes it possible to fabricate the large TFT

Oxide Thin Film Transistors, First Edition. Yue Kuo, Hideo Hosono, Michael S. Shur, and Jin Jang.

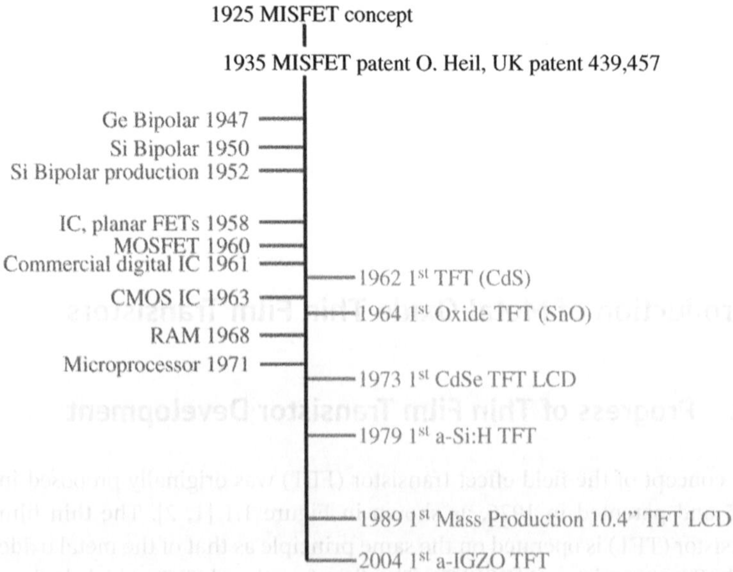

Figure 1.1 History of TFT and IC development. Source: Kuo et al. [1]/John Wiley & Sons.

array over a low-cost glass or even a plastic substrate. The gate dielectric and channel layers can be deposited by one-pump-down, which warrants the high performance and reliability of the TFT. The 10.4″ direct-view TFT LCD was included in the IBM ThinkPad PC in 1989. Other a-Si:H TFT-based products, such as X-ray medical imagers, gas, pH, photo, and magnetic sensors, were published in the literature [9]. Commercial 98″ TFT LCD TVs were available on the market in 2023 [10]. Glass substrates as large as 3 m by 3 m have been used in the 10.5th generation production equipment. The TFT LCD industry became the second largest electronic industry after the IC industry.

The main disadvantage of the a-Si:H TFT is the low field effect mobility (μ_{eff}), i.e., typically less than 1 $cm^2/V\,s$. which limits its applications in high-speed or high driving-current products, such as the display driver circuit or the organic light-emitting diode (OLED). On the other hand, the polycrystalline silicon (poly-Si) thin film has been used as the gate electrode in MOSFETs since 1970s. It is natural for researchers to develop poly-Si TFTs from the early stage. The poly-Si TFT can have a mobility 2 orders of magnitude higher than that of the a-Si:H [11]. However, it is difficult to mass-produce poly-Si TFT arrays on a large-area substrate at a low temperature due to manufacturing reasons. The high-mobility poly-Si TFT is commonly made of high-temperature deposited poly-Si film or thermally annealed a-Si film, which is incompatible with the thermal stability of commercial display glass. The low thermal-budget laser annealing method, which provides the best poly-Si characteristics, suffers from low throughput. Therefore, currently, most poly-Si TFTs are used in small-sized displays.

Despite the early report of the high mobility SnO_2 TFT, i.e., $\mu_{eff} = 70\,cm^2/V\,s$, in 1964 [6], the metal oxide TFT was never used in commercial products until the report of the amorphous indium gallium zinc oxide (a-IGZO) TFT by the Hosono's group in 2004 [12]. The renewed interest in industry and academia came from the possibility of mass production of the high μ_{eff} at a low temperature over a large-area substrate. The original a-IGZO TFT was made of the pulsed laser ablation deposition method at room temperature. The original large off-current (I_{off}) was later reduced to $<10^{-14}$ A [13]. Also, the complete TFT could be fabricated by the sputtering or PECVD process, which is suitable for the large-area, low-temperature, and high-throughput production requirements.

1.2 Progress of Metal Oxide Thin Film Transistor Development

Figure 1.2 shows a short summary of some important events related to metal oxide TFTs. For the rest of this book, the term "metal oxide TFT" will be shortened as "oxide TFT." Shockley and Pearson first reported that the p-type semiconductor Cu_2O could be used as a channel material for the FET 1948 [14]. The result remained at the current field modulation level and almost unchanged to date. Although several Cu^+-based p-type oxide semiconductors, such as $CuAlO_2$, have been reported, no good performance TFT has been reported so far [15]. In 1954, semiconducting oxide glasses based on V_2O_x-P_2O_5 were reported by a group at Sheffield University [16]. This is the first report on the electronic conducting (bulk) glass, but the conduction is due to variable-range hopping between localized V^{4+} and V^{5+}. Thus, this type of oxide semiconductor has very small mobility and cannot be used as the TFT channel material. Since then, it is well-known that transition metal oxides exhibit semiconducting properties. Transition metal cations have an open shell structure in their d-orbitals and give visible absorption originating from a d-d transition. Since these vacant d-levels give a large density of states (DOS) in the gap, it is hard to shift Fermi level (E_F) significantly exceeding these DOS. This is the reason why transition metal oxide semiconductors do not work as TFT channel materials. Requirements on TFT channel semiconductor materials are much more severe than those for p-n junction formation because the E_F has to shift to beyond the valence

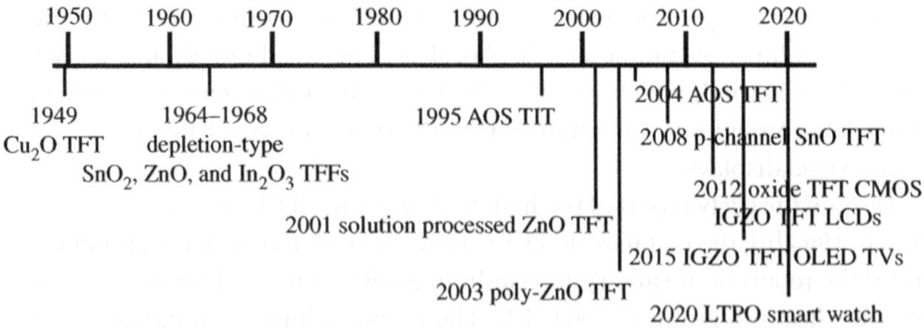

Figure 1.2 Examples of important events related to metal oxide TFTs.

band maximum (VBM) or the conduction band minimum (CBM) upon the application of a gate voltage. High sensitivity of Cu^+ to oxidation to Cu^{2+} at the surface and grain boundary [17] prevents the gate-bias-induced band conduction notwithstanding Cu_2O thin films have high Hall mobility of ~90 $cm^2/V\,s$ [17].

Since the first report of the SnO_2 TFT by Klasens and Koelmans at Philips Research Laboratories in 1964 [6], there have been many studies on oxide TFTs. In 1972, a study of the optical and electrical properties of Cd_2SnO_4 revealed its semiconductor properties [18]. In 1996, Hosono et al. [19] described the electrical properties of three transparent conducting amorphous oxides: a-$AgSbO_3$, Cd_2GeO_4, and Cd_2PbO_4 having an electron mobility of ~10 $cm^2/V\,s$. This discovery was followed by the report on crystalline transparent oxide semiconductors [20] and by the report on the room-temperature fabrication of transparent flexible TFTs using amorphous oxide semiconductors [12]. In 2003, Hoffman et al. [21] reported on the ZnO-based transparent thin-film transistor.

Oxide TFT research was revisited extensively by several groups since 2000. A motivation was to develop TFTs with higher mobility and facile fabricability for next-generation flat panel displays (FPDs), such as OLEDs. Among them, noteworthy is a paper in 2001 by Ohya et al., which is the first report of a non-vacuum, solution process in preparing ZnO-TFTs, i.e., drop-coating of $Zn(CH_3COO)_2$ solution and subsequent heating in air [22]. Since high chemical stability in an ambient atmosphere is one of the material advantages of oxide semiconductors, this approach utilizing this intrinsic nature of oxides became a milestone in the fabrication of solution-derived oxide TFTs. Many papers on ZnO-TFTs deposited by sputtering or pulsed laser deposition were reported but serious issues, such as poor reproducibility and large hysteresis, arising mainly from the complex grain boundary nature, were pointed out. The design concept of transparent amorphous oxide semiconductors (TAOS) with large electron

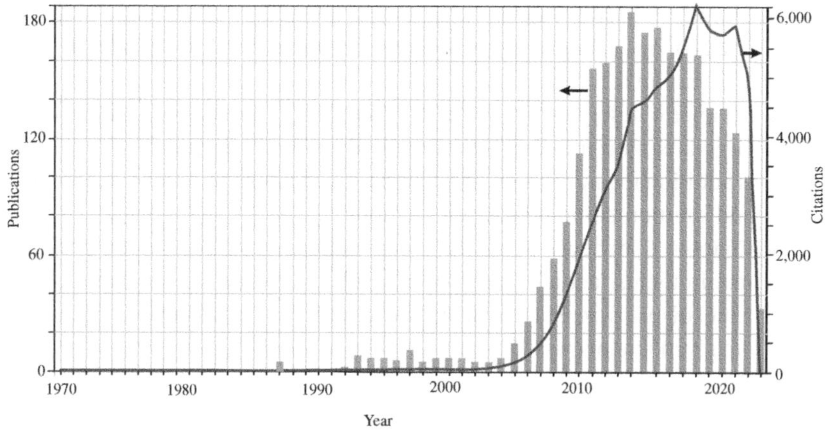

Figure 1.3 Number and citation of publication with title containing "oxide" and "thin film transistors". Source: Data were taken from Web of Science.

mobility was proposed by Hosono et al. [23] in 1996 along with several demonstrative materials. Although amorphous InO_x had large electron mobility and high transparency, the performance of a-InO_x TFTs was insufficient due to high carrier concentration [24]. A TAOS material, i.e., IGZO, appropriate for TFTs was found [12]. First, IGZO TFT was fabricated using epitaxial thin films of IGZO in 2003, exhibiting excellent performance, mobility of ~80 $cm^2/V\,s$ and on- and off-current ratio (I_{on}/I_{off}) of 10^6 [25]. Subsequently, amorphous IGZO TFT fabricated on plastic substrates with mobility of ~8 $cm^2/V\,s$ was reported [12].

Figure 1.3 shows the number of SCI journal papers with titles containing "oxide thin film transistor" and their citations from 1970 to 2023. The number of oxide TFT papers and their citation started to boost in 2004. This distinct trend suggests these two papers triggered much attention of TFT researchers to oxide semiconductors and this excitement, primarily on IGZO, has continued to date. Many monographs and review papers on oxide TFTs have been published [13, 15, 26–52].

1.3 Layout of Book Chapters

The arrangement of chapters follows the concept of materials, devices, processes, and applications.

- Chapter 2 discusses fundamental electronic properties and material properties of oxide semiconductors. The emphasis is on issues related to the material requirement in preparing high-performance oxide TFTs.

- Chapter 3 discusses device physics and the design of oxide TFTs. Special issues on modeling, tera-hertz operation, and reliability of oxide TFTs are analyzed in this chapter.
- Chapter 4 describes the oxide TFT fabrication processes considering the large-area thin film deposition methods and transistor structures. There is additional emphasis on the fabrication of emerging flexible oxide TFTs and solution-based processing methods.
- Chapter 5 reviews oxide TFT applications. In addition to principal requirements, actual examples of applications on various types of flat panel displays, sensors, imagers, and circuits as well as flexible electronics are given and discussed.

References

1 Kuo, Y., Jang, J., and Nathan, A. (2023). Chapter10, The thin film transistor and emergence of large area, flexible electronics and beyond. In: *IEEE Special Anniversary Commemorative Volume to Celebrate The 75th Year Invention of Transistor*, 263–272. New York: Wiley https://doi.org/10.1002/9781394202478.ch21.

2 Lilienfeld, J.E. (1926). Method and apparatus for controlling electric currents. U.S. Patent 1745175A.

3 Weimer, P.K. (1962). The TFT a new thin-film transistor. *Proceedings of the IRE* 50 (6): 1462–1469. https://doi.org/10.1109/JRPROC.1962.288190.

4 Ticle, A.C. (1969). *Thin-Film Transistors: A New Approach to Microelectronics*. New York: Wiley.

5 Kuo, Y. (1989). Thin film technologies in active matrix addressing system of LCD. *SPIE Proceedings of Display System Optics II* 1117: 114–122. https://doi.org/10.1117/12.960928.

6 Klasens, H.A. and Koelmans, H. (1964). A tin oxide field-effect transistor. *Solid-State Electronics* 7 (9): 701–702. https://doi.org/10.1016%2F0038-1101%2864%2990057-7.

7 Brody, T.P., Asars, J.A., and Dixon, G.D. (1973). A 6 × 6 Inch 20 lines-per-inch liquid crystal display panel. *IEEE Transactions on Electron Devices ED-20* (11): 995–1001. https://doi.org/10.1109/T-ED.1973.17780.

8 le Comber, P.G., Spear, W.E., and Ghaith, A. (1979). Amorphous-silicon field-effect device and possible application. *Electronics Letters* 15 (6): 179–181. https://doi.org/10.1049/el:19790126.

9 Kuo, Y. (2013). Thin film transistor technology – past, present, and future. *ECS Interface* 22: 55–61.

10 [Online]. "TCL 98" CLASS XL COLLECTION 4K UHD QLED DOLBY VISION HDR SMART GOOGLE TV – 98R754" https://www.tcl.com/us/en/products/home-theater/xl-collection/98-class-4k-qled-hdr-smart-google-tv-98R754 (accessed 31 July 2024).

11 Sameshima, T. (2004). Chapter 5, Poly-Si TFTs by laser crystallization methods. In: *Polycrystalline Silicon Thin Film Transistors* (ed. Y. Kuo), 176–219. Boston: Kluwer.

12 Nomura, K., Ohta, H., Takagi, A. et al. (2004). Room-temperature fabrication of transparent flexible thin-film transistors using amorphous oxide semiconductors. *Nature* 432: 488–492. https://doi.org/10.1038/nature03090.

13 Kamiya, T., Nomura, K., and Hosono, H. (2010). Present status of amorphous In–Ga–Zn–O thin-film transistors. *Science and Technology of Advanced Matters* 11 (4): 044305. https://doi.org/10.1088/1468-6996/11/4/044305.

14 Shockley, W. and Pearson, G.L. (1948). Modulation of conductance of thin films of semiconductors by surface charges. *Physical Review* 74 (2): 232–233.

15 Al-Jawhari, H.A. (2015). A review of recent advances in transparent p-type Cu_2O-based thin film transistors. *Materials Science in Semiconductor Processing* 40: 241–252.

16 Denton, E.P., Rawson, H., and Stanworth, J.E. (1954). Vanadate glasses. *Nature* 173 (4413): 1030–1032.

17 Deuermeier, J., Wardenga, H.F., Morasch, J. et al. (2016). Highly conductive grain boundaries in copper oxide thin films. *Journal of Applied Physics* 119 (23): 235303. https://doi.org/10.1063/1.4954002.

18 Nozik, A.J. (1972). Optical and electrical properties of Cd_2 SnO_4: a defect semiconductor. *Physical Review B* 6 (2): 453–459. https://doi.org/10.1103/PhysRevB.6.453.

19 Hosono, H., Yasukawa, M., and Kawazoe, H. (1996). Novel oxide amorphous semiconductors: Transparent conducting amorphous oxides. *Journal of Non-Crystalline Solids* 203: 334–344. https://doi.org/10.1016/0022-3093(96)00367-5.

20 Matsuzaki, K., Nomura, K., Yanagi, H. et al. (2008). Epitaxial growth of high mobility Cu_2O thin films and application to p-channel thin film transistor. *Applied Physics Letters* 93 (20): 202107.

21 Hoffman, R.L., Norris, B.J., and Wager, J.F. (2003). ZnO-based transparent thin-film transistors. *Applied Physics Letters* 82: 733–735. https://doi.org/10.1063/1.1542677.

22 Ohya, Y., Niwa, T., Ban, T. et al. (2001). Thin film transistor of ZnO fabricated by chemical solution deposition. *Japanese Journal of Applied Physics* 40 (1R): 297–298.

23 Hosono, H., Kikuchi, N., Ueda, N. et al. (1996). Working hypothesis to explore novel wide band gap electrically conducting amorphous oxides and examples. *Journal of Non-Crystalline Solids* 198: 165–169.

24 Adkins, C.I., Hussain, T., and Ahmad, N. (1993). Field-effect measurements of carrier mobilities in transparent conducting films of amorphous indium oxide. *Journal of Physics. Condensed Matter* 5 (36): 6647–6652.

25 Nomura, K., Ohta, H., Ueda, K. et al. (2003). Thin-film transistor fabricated in single-crystalline transparent oxide semiconductor. *Science* 300 (5623): 1269–1272. https://doi.org/10.1126/science.1083212.

26 Higashiwaki, M. and Fujita, S. (ed.) (2020). *Gallium Oxide: Materials Properties, Crystal Growth, and Devices*, vol. 293. Springer Nature Fundamentals of Ga_2O_3 and their application to power electronics are summarized.

27 Ellmer, K., Klein, A., and Rech, B. (ed.) (2007). *Transparent Conductive Zinc Oxide: Basics and Applications in Thin Film Solar Cells*. Springer. Fundamentals of ZnO as the transparent conductors are summarized along with solar cell applications.

28 Ginley, D.S., Hosono, H., and Perkins, J.D. (ed.) (2010). *Handbook of Transparent Conductors*, 1–25. Boston, MA: Springer, Fundamentals and advances of transparent conductive oxides are described.

29 Wager, J.F., Keszler, D.A., and Presley, R.E. (2008). *Transparent Electronics*, vol. 112. New York: Springer Research on transparent oxide semiconductors was well reviewed up to 2007 covering materials, Devices (TFTs) and circuits.

30 Facchetti, A. and Marks, T.J. (2010). *Transparent Electronics. From Synthesis to Applications*. Wiley Transparent electronic related materials (oxide thin films, nanotubes, organic semiconductors) are described along with their device applications including flexible devices.

31 Barquinha, P., Martins, R., Pereira, L. et al. (2012). *Transparent Oxide Electronics: from Materials to Devices*. Wiley. Wide gap oxide semiconductors and insulators for electronic devices including paper electronics are described.

32 Hosono, H. and Kumomi, H. (ed.) (2022). *Amorphous Oxide Semiconductors: IGZO and Related Materials for Display and Memory*. Wiley. Fundamentals and TFT application of amorphous oxide semiconductors are summarized comprehensively.

33 Kuo, Y. and Furuta, M. (Guest Eds.) (2014). Focused issue on oxide thin film transistors. *Journal of Solid State Science and Technology* 3 (9): Q3001–Q3094.

34 Hosono, H. (2006). Ionic amorphous oxide semiconductors: material design, carrier transport, and device application. *Journal of Non-Crystalline Solids* 352 (9–20): 851–858.

35 Street, R.A. (2009). Thin-film transistors. *Advanced Materials* 21 (20): 2007–2022.

36 Fortunato, E., Barquinha, P., and Martins, R. (2012). Oxide semiconductor thin-film transistors: a review of recent advances. *Advanced Materials* 24 (22): 2945–2986.

37 Nathan, A., Lee, S., Jeon, S. et al. (2014). Amorphous oxide semiconductor TFTs for displays and imaging. *Journal of Display Technology* 10 (11): 917–927.

38 Yu, X., Marks, T.J., and Facchetti, A. (2016). Metal oxides for optoelectronic applications. *Nature Materials* 15 (4): 383–396.

39 Petti, L., Münzenrieder, N., Vogt, C. et al. (2016). Metal oxide semiconductor thin-film transistors for flexible electronics. *Applied Physics Reviews* 3 (2): 021303. https://doi.org/10.1063/1.4953034.

40 Wang, Z., Nayak, P.K., Caraveo-Frescas, J.A. et al. (2016). Recent developments in p-Type oxide semiconductor materials and devices. *Advanced Materials* 28 (20): 3831–3892.

41 Wager, J.F. and Kuhn, K. (2017). Device physics modeling of surfaces and interfaces from an induced gap state perspective. *Critical Reviews in Solid State and Materials Sciences* 42 (5): 373–415.

42 Sheng, J., Lee, J.H., Choi, W.H. et al. (2018). Atomic layer deposition for oxide semiconductor thin film transistors: advances in research and development. *Journal of Vacuum Science & Technology A* 36 (6): 060801.

43 Ide, K., Nomura, K., Hosono, H. et al. (2019). Electronic defects in amorphous oxide semiconductors: a review. *Physica Status Solidi A* 216 (5): 1800372.

44 Lee, S.Y. (2020). Comprehensive review on amorphous oxide semiconductor thin film transistor. *Transactions on Electrical and Electronic Materials* 21: 235–248.

45 Park, J.W., Kang, B.H., and Kim, H.J. (2020). A review of low-temperature solution-processed metal oxide thin-film transistors for flexible electronics. *Advanced Functional Materials* 30 (20): 1904632.

46 Jenifer, K., Arulkumar, S., Parthiban, S. et al. (2020). A review on the recent advancements in tin oxide-based thin-film transistors for large-area electronics. *Journal of Electronic Materials* 49: 7098–7111.

47 Ji, D., Jang, J., Park, J.H. et al. (2021). Recent progress in the development of backplane thin film transistors for information displays. *Journal of Information Display* 22 (1): 1–11.

48 Nomura, K. (2021). Recent progress of oxide-TFT-based inverter technology. *Journal of Information Display* 22 (4): 211–229.

49 Yan, A., Wang, C., Yan, J. et al. (2023). Thin-film transistors for integrated circuits: fundamentals and recent progress. *Advanced Functional Materials* 34 (3): 2304409. https://doi.org/10.1002/adfm.202304409.

50 Kim, T., Choi, C.H., Hur, J.S. et al. (2022). Progress, challenges, and opportunities in oxide semiconductor devices: a key building block for applications ranging from display backplanes to 3D integrated semiconductor chips. *Advanced Materials* 35 (43): 2204663. https://doi.org/10.1002/adma.202204663.

51 Chen, F., Zhang, M., Wan, Y. et al. (2023). Advances in mobility enhancement of ITZO thin-film transistors: a review. *Journal of Semiconductors* 44 (9): 091602.

52 Charnas, A., Zhang, Z., Lin, Z. et al. (2023). Extremely thin amorphous indium oxide transistors. *Advanced Materials* 36: 2304044.

2

Semiconductor Materials in Oxide Thin Film Transistor

2.1 Introduction

Oxide semiconductors have characteristics rather different from conventional semiconductors based on diamond structure like Si. This chapter introducing the outline of oxide semiconductor materials is composed of two parts; the first part (Section 2.2) describes the fundamentals of oxide semiconductors and their TFTs in comparison with Si from views of chemical bonding and electronic structure and shows a comprehensive view of doping ability and p/n orientation from band alignment. Besides, the approach to p-type materials and p/n control are explained with fundamental material properties. Amorphous hydrogenated Si (a-Si:H) had been exclusively used until ~2012 as the backplane TFTs to drive pixels of flat panel displays. However, oxide semiconductor TFT is becoming a strong alternative due to their high mobility, low off-current, and facile fabrication by conventional sputtering method. In particular, TAOS semiconductors, such as In–Ga–Zn–O (IGZO), are widely applied as the backplane for high resolution and energy-saving LCDs and large-sized OLED TVs and are examining the application to semiconductor memory coupled with Si utilizing very small off current. Given such background, the second part (Section 2.3) highlights the material features of amorphous oxide semiconductors including IGZO and their application to TFT channels. Finally, technical and fundamental issues in this area are described in Section 2.4.

2.2 Electronic and Crystal Structure of Wide-gap Oxides

Wide-gap oxide materials such as SiO_2 and Al_2O_3 are known as the main ingredients of ceramics such as porcelains and glasses. Since they have

Oxide Thin Film Transistors, First Edition. Yue Kuo, Hideo Hosono, Michael S. Shur, and Jin Jang.

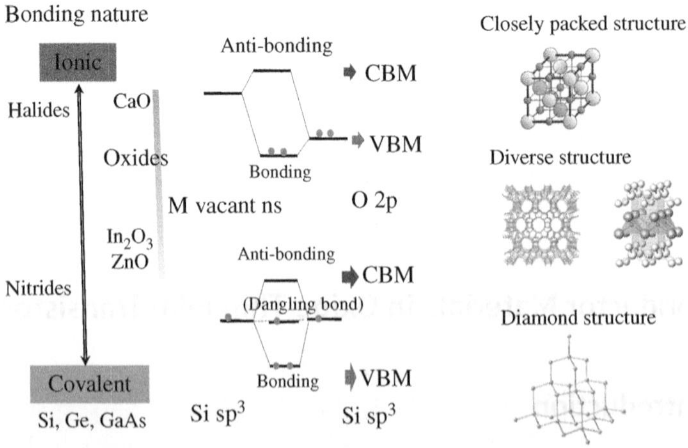

Figure 2.1 Chemical bonding and crystal structure. Diversity of crystal structure is a characteristic of metal oxides reflecting varied bonding nature with cationic species.

bandgaps much greater than 3 eV, these materials are optically transparent and electrically insulating. Transparent oxides are connected by chemical bonding with a rather ionic nature, in contrast to typical semiconductors such as Si and GaAs, which are covalently bonded.

Figure 2.1 shows outlines of the electronic and crystalline structures of transparent oxides and a typical semiconductor. The latter has a diamond-like crystal structure comprising tetrahedrons as the building blocks similar to Si crystals and the structure of compound semiconductors is derived from this simple structure. Two sp^3 hybrid orbitals of adjacent atoms overlap to form bonding and anti-bonding orbitals depending on the difference in orbital phases, leading to the valence and conduction bands, respectively. The splitting of the energy forms a bandgap. Hence, the contribution of the atomic orbitals is similar for the two bands. In contrast, ionic crystalline structures can generally be explained as those in which large anions are densely packed and the spaces around are occupied by cations. Most oxides have strongly ionic bonding but the contribution of covalent bonding is not negligible. Therefore, the crystal structure of oxides is not merely an agglomeration of spheres but includes various forms, such as a crystalline structure comprising a continuous set of cage structures and a layered structure, because of the bonding anisotropy, which is unique to covalently bonded crystals, in addition to the covalent properties. Moreover, there are a huge number of oxide types because almost all the metal elements in the periodic table can become cations to form oxides. The diverse range of types and structures is a distinctive feature of oxides. The selection of materials suitable for deriving desired functionalities from among the various types of oxide is a major theme of material research attracting interest.

For highly ionic materials, the contribution of anions is dominant at the valence band maximum (VBM), whereas that of cations is dominant at the conduction band minimum (CBM). For the oxides of typical metal elements, the VBM is mainly composed of oxygen 2p orbitals and the CBM is mainly composed of empty s orbitals of a metal cation, meaning that the characteristics of the material are strongly manifested at the CBM. For the oxides of transition-metal elements with partially occupied d orbitals, the occupied and unoccupied d orbitals are at deep levels and thus the bandgap is smaller than 3 eV, resulting in colored materials.

2.2.1 Requirements for Semiconductors as Excellent TFT Channel Materials

There are several requirements for TFT channel semiconductors. First is a well-controllable Fermi level by gate voltage. This requirement means to have less trap state density between the E_F and band edge (VBM for p-channel TFT, CBM for n-channel). This requirement rules out oxide semiconductors of transition metals with open-shell structures and conventional amorphous semiconductors including H-free a-Si. The second is surface stability. Oxide semiconductors based on Cu^+ such as Cu_2O do not work as TFT channel notwithstanding the thin film has large Hall mobility, $\sim$100 cm^2/V s, and appropriate carrier concentration (10^{13-14} cm^{-3}). The cause is due to the easy oxidation of Cu from +1 to +2 that occurs on the surface upon exposure to the ambient atmosphere or the negative gate bias. The third requirement is the low carrier density. Since TFTs take on the role of voltage-driven switching in circuits, a large ON/OFF current ratio (I_{on}/I_{off}) is demanded. The low carrier concentration leads to low I_{off} in TFTs, which results in low power consumption. Transparent conductive oxides used for transparent contacts such as In_2O_3:Sn (ITO) and ZnO:Al (AZO) exhibit metallic conduction arising from large carrier concentrations ($\sim 10^{20}$ cm^{-3}), but they are almost impossible to control conductivity over a wide range by gating. The fourth is the easy formation of low contact resistance between channel and electrode materials. This is particularly important for TFTs for memory application because the ratio of contact resistance to channel resistance is large in down-sized devices. The last is the gate insulator which can form a good interface with channel semiconductors.

One may understand from these requirements described above that semiconductors suitable for TFT channels are rather restricted among general semiconductors.

2.2.2 Electronic Structure of N- and P-type Oxide Semiconductors

As described in Figure 2.1, non-transition metal oxides have an almost common electronic structure, less dispersive valence band maximum primarily composed of localized O2p orbitals and dispersed conduction band to which metal vacant s-orbitals contribute largely. Since almost all metal cations may form oxides, n-type oxide semiconductors can be designed by selecting metal cations with spatially spread s orbitals constituting the CBM and a crystalline structure with a smaller cation–cation separation. Figure 2.2 shows a band structure of SnO_2, which is a typical wide-gap n-type semiconductor [1]. This material has a rutile structure (TiO_2-type) composed of edge-shared SnO_6 octahedra. Sn^{4+} 5s orbitals primarily form the CBM and their direct overlap is rather large irrespective of intervening O2p.

This situation comes from the large spatial spread of Sn5s orbital relative to Sn–Sn separation in edge-sharing structure as shown in Figure 2.3a. The large spread of metal s-orbital relative to metal–metal separation is a requirement for n-type oxide semiconductors. Such a metal cation corresponds to post-transition in the periodic table. Figure 2.3b shows the representative n-type semi-conductive oxides. Since solid solutions among each oxide exhibit n-type conduction generally, the compositional territory for n-type oxide semiconductors is within this polyhedron, i.e., there are

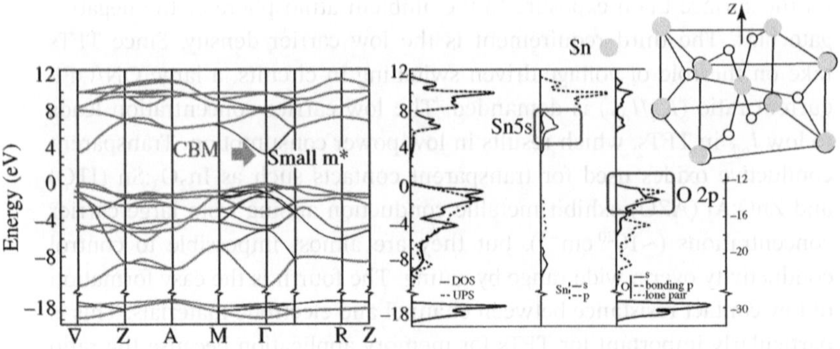

Figure 2.2 Band structure and crystal structure of SnO_2 (rutile-type). Source: Robertson [1]/IOP Publishing.

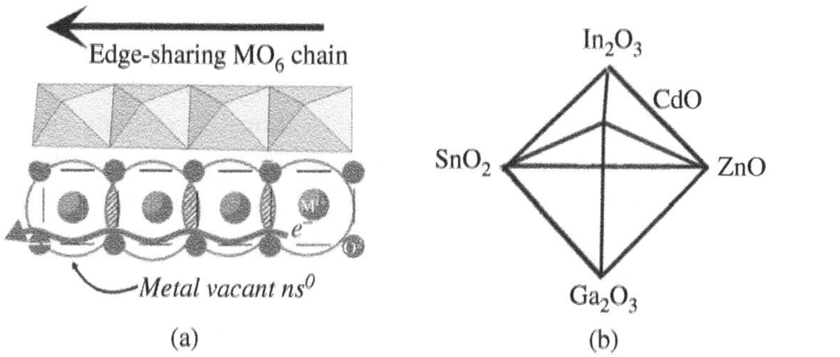

Figure 2.3 (a) Edge-sharing and (b) crystal structures suitable for n-type oxide semiconductors.

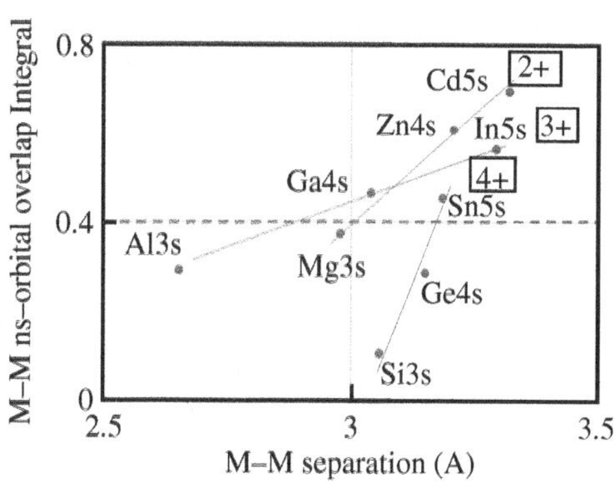

Figure 2.4 Calculated overlap integral between metal ns orbitals and metal–metal cation separation in simple oxides. Oxides; MgO, Al_2O_3, SiO_2, ZnO, β-Ga_2O_3, GeO_2, CdO, In_2O_3, and SnO_2. Slater-type orbital function was used. Source: Orita et al. [2]/Taylor and Francis Group.

a variety of possible semiconductors. Appropriate post transition metal cations for n-type oxide semiconductors are 5, Sn^{4+}, In^{3+}, Ga^{3+}, Cd^{2+}, and Zn^{2+}.

The choice of these metal cations is justified by the relation between the overlap of neighboring metal vacant ns-orbitals and the separation of these metal cations in simple oxides shown in Figure 2.4 [2]. It is evident from the figure that the overlap of these neighboring s-orbitals intervening oxygen is larger than that of non-post transition metal cations irrespective of much larger metal–metal separation. This finding strongly supports percolated vacant ns-orbitals of these cations form the conduction band minimum with large dispersion (small effective mass), leading to a large electron affinity (>~4 eV) for which electron doping is possible as will be described in Section 2.2.3.

On the other hand, no guidelines for designing p-type semiconductors were reported until 1997. Needless to say, pn junctions are the origin of various semiconductor functions; therefore, high-quality p-type wide-gap oxide semiconductors are essential for extension of oxide electronics. In wide-gap oxides, the VBM, which serves as the conduction path of holes, is mainly composed of oxygen 2p orbitals with highly localized nature and the contribution of the orbitals of metal cations is generally small. Therefore, the VBM is little dispersed (i.e., the effective mass of holes is large), and the energy level is deep. This is why p-type transparent oxide semiconductors are difficult to realize. Resolving this problem is a key to realizing p-type transparent oxide semiconductors.

Three proposed approaches are summarized below:

(1) Oxides of transition-metal cations with electronic configuration of $(n-1)d^{10}ns^0$ ($n = 4$ or 5) [3]
Although transition-metal cations have d-orbitals with energy levels close to that of oxygen 2p orbitals, most of them absorb visible light owing to a d–d transition. Therefore, oxides of cations with closed-shell d orbitals, such as Cu^+ and Ag^+, are considered to have the potential to exhibit p-type conductivity. As shown in Figure 2.5, for such oxides, the antibonding orbital component of the metal-filled d-orbitals and oxygen 2p orbitals constitutes the VBM, and the holes doped into the VBM are delocalized because O2p forms a covalent bonding with the M d-orbitals. A typical example is copper delafossite $CuMO_2$ (M = Al^{3+}, Ga^{3+}, In^{3+}) with dumbbell-type O—Cu—O bonds as the building block [3, 4]. Please note that Cu^+–O–Cu^+ linear unit constitutes p-type semiconductor Cu_2O but the layer structure of delafossite widens

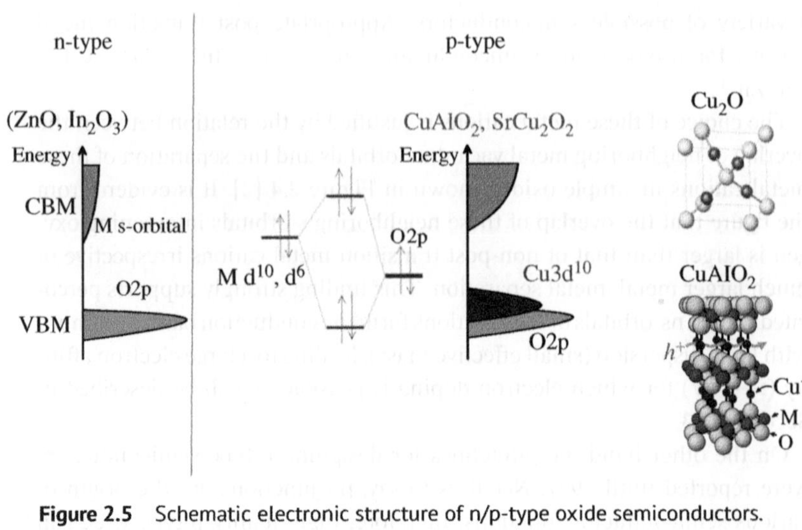

Figure 2.5 Schematic electronic structure of n/p-type oxide semiconductors.

the band gap compared with Cu_2O suppressing the VBM dispersion, leading to a wide gap p-type oxide semiconductor. Although several Cu^+-based p-type transparent semiconductors have been found to date, no good TFT operation has been reported to date like the Cu_2O case. The high concentration of hole traps at the surface associated with oxidation of Cu^+ would be the most plausible cause of these results. While Ag^+-based oxides have similar band structure to the cuprous ion-based, the high sensitivity of Ag^+ to light, i.e., reduction to Ag^0, hinders the detailed research.

(2) Oxides of metal cations with electronic configuration of ns^2
Cations with an electronic configuration of the form ns^2 have lone pairs similar to those of anions. For oxides of such metal cations, the VBM is mostly occupied by these s orbitals, as shown in Figure 2.6. Lone pairs of the spatially dispersed s orbitals, which overlap with the s orbitals of adjacent cations via the oxygen 2p, form a largely dispersed band above

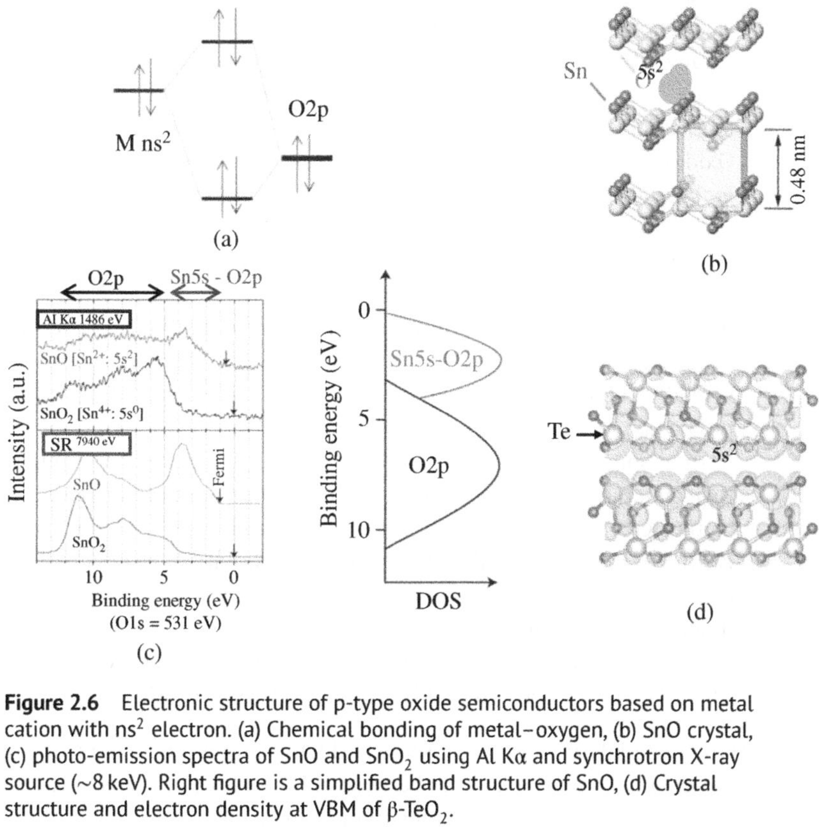

Figure 2.6 Electronic structure of p-type oxide semiconductors based on metal cation with ns^2 electron. (a) Chemical bonding of metal–oxygen, (b) SnO crystal, (c) photo-emission spectra of SnO and SnO_2 using Al Kα and synchrotron X-ray source (~8 keV). Right figure is a simplified band structure of SnO, (d) Crystal structure and electron density at VBM of β-TeO_2.

the oxygen 2p band. Oxides of Sn^{2+} with an electronic configuration of $5s^2$ are typical and exhibit p-type conductivity in Hall effect measurement. In addition, the first oxide p-channel TFT was realized using SnO [5]. Realization of C-MOS based on an oxide semiconductor was a long-standing issue in oxide electronics. This objective was first attained in 2011 using SnO [6]. Although much effort has been paid for performance improvement [7], the C-MOS performance is still insufficient for application. Low-temperature processing as well as improvement of mobility is strongly required for display application.

It has been reported that oxides of Pb^{2+} and Bi^{3+}, both with an electronic configuration of $6s^2$, are not favorable for enhancement of hole transport because the energy level of 6s electrons is much deeper than the VBM mainly composed of O2p due to relativistic effect [8]. Although oxides of Sb^{3+} with an electronic configuration of $5s^2$, similar to that of Sn^{2+}, are expected to exhibit p-type conductivity, no examples of such oxides have been reported to date. Recently, β-TeO_2 with 2D structure was reported to work as p-channel TFTs with high mobility [9]. This oxide of Te^{4+} has the same electronic configuration as Sn^{2+} and Sb^{3+}. Since Te^{4+} is easily reduced to Te, caution would be required for fabrication.

(3) Oxides of metal cations with electronic configuration of nd^6

For oxides of Rh^{3+} and Ir^{3+}, both with an electronic configuration of $4d^6$ or $5d^6$, these cations stabilize in the low-spin-state octahedral configuration, where electrons occupy the three orbitals, i.e., d_{xy}, d_{yz}, and d_{zx}. These 4d and 5d orbitals are spatially spread, and the state in which two electrons occupy each of the three d orbitals is like that in (2), i.e., they may be regarded as pseudo s-orbital with large spatial spread. When doped with holes, these oxides are expected to exhibit p-type conductivity. A typical example is $ZnRh_2O_4$ [10], which has a normal spinel structure. Similar to the case of ZnO, the Zn^{2+} ions in this material are tetrahedrally coordinated but are not continuously connected, thus this material does not exhibit n-type conductivity. $ZnRh_2O_4$ is the only oxide that is known to exhibit p-type conductivity even in the amorphous state [11], and has been reported to form pn diodes on plastic substrates when combined with n-type TAOS [12]. Later, p-type conductivity and pn-junction with ZnO were reported for $ZnCo_2O_4$ as an extension of this series to $3d^6$ system [13].

These materials work as p-type semiconductors for pn-junction but do not do as p-channel material in TFTs because of high concentrations of vacant d levels which serve as hole trapping in the band gap.

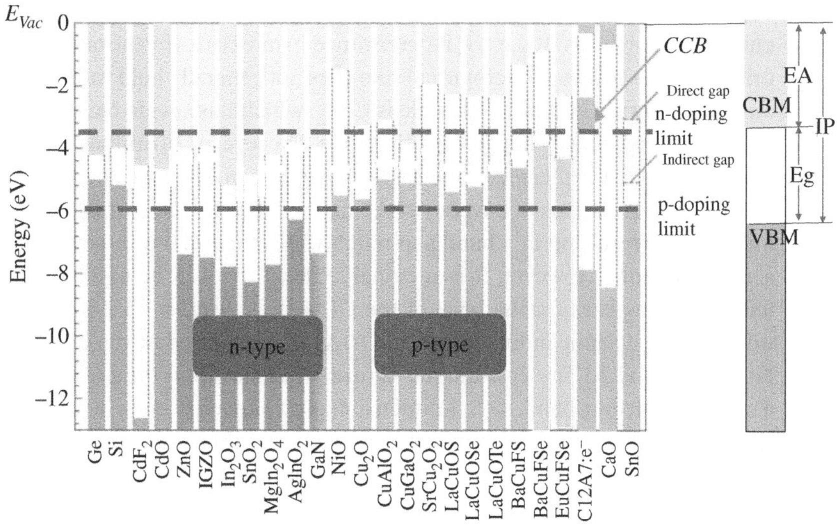

Figure 2.7 Band alignment of various semiconductors. Source: Hosono [14]/IOP Publishing. Dotted areas indicate empirical doping windows for carrier generation. IP: ionization potential, EA: electron affinity.

2.2.3 Doping Limit and Band Alignment

Carrier dopability and p/n orientation are the major concerns in semiconductor science and technology. Here these issues are considered on the basis of the band energy lineup of semiconductor materials. The levels of the CBM and VBM for various materials are lined up with respect to the vacuum level (E_{vac}). Figure 2.7 shows a band lineup of oxide semiconductors and relevant semiconductors. The following findings were obtained from the band lineup [14]:

(1) The CBM of oxides that can be doped with electrons is deeper than ~ -4 eV from the E_{vac}. This means electron affinity (EA) of $>\sim 4$ eV is favorable for n-type semiconductors.
(2) The VBM of oxides that can be doped with holes is shallower than ~ -6 eV, i.e., ionization potential (IP) of smaller than ~ 6 eV is favorable for p-type semiconductors.
(3) Representative ambipolar semiconductors Si and Ge have both EA and IP located near the center of the empirical doping window (shown in a dotted rectangle).

These findings mean that the pn orientation is determined by the stability of carriers. The empirical rule that wide-gap materials are difficult to dope with carriers mostly holds *when the type of anion is fixed.* Since the energy level of the VBM, i.e., IP, is determined in most cases by the outermost orbitals of the anions, this empirical rule does not generally hold valid when anions are varied. A typical example is CdF_2, which has a wide band gap of ~7 eV but can be easily doped with electrons. The level of 2p orbitals of fluorine, which has a higher electronegativity than oxygen, is 4–5 eV deeper than the level of 2p orbitals of oxygen, lowering the VBM by the corresponding amount and resulting in a band gap much wider than that of n-type oxide semiconductors. However, it is reasonable that CdF_2 can be easily doped with electrons because the level of the CBM is as deep as −5 eV. A quite unique n-type semiconductor is $12CaO{\cdot}7Al_2O_3$ composed of typical insulators CaO and Al_2O_3. This unique nature originates from the presence of a new conduction band, i.e., 3-dimensionally connected cages with positive charge by sharing a monoxide layer constitutes this unique band called "cage conduction band" (CCB) [15]. Since this CCB is located below the conventional conduction band primarily composed of Ca 4s-orbitals and the CCB bottom is close to −4 eV from the E_{vac}, we may understand successful electron-doping to $12CaO{\cdot}7Al_2O_3$ on the basis of the band alignment.

SnO is the first oxide semiconductor whose conduction can be controlled between p- and n-type conduction and has been confirmed to exhibit ambipolar operation in the TFT structure, i.e., the p- and n-channels are switched depending on the sign of the gate voltage [6]. The direct gap of SnO is 2.7 eV, from which the CBM (direct) is estimated to be positioned at −3.1 eV relative to E_{vac}, indicating difficulty in electron doping. However, a band calculation reveals the existence of an indirect gap, which is experimentally estimated to be 0.7 eV from optical absorption measurement. Thus, the CBM (corresponding to indirect gap) is positioned at −4.4 eV relative to E_{vac}, indicating the possibility of n-type doping. Actually, we carried out an experiment of impurity doping by replacement and found that the carrier polarity was converted from p-type to n-type depending on the impurity concentration [16].

The band-lineup-based findings (1) and (2) were obtained by a phenomenological approach based on currently available research results and do not exclude the possibility that the limits can be exceeded by methods different from those conventionally used, such as doping in a nonequilibrium process. Moreover, the control of the conduction type depends on not only the stability of doped carriers but also the ease of occurrence of defects that generate opposite-polarity carriers. For example, CdO is a material that can potentially be controlled between p- and n-type conduction according to Figure 2.7. However, only n-type conduction has been realized thus far. Although the electron carrier concentration can be considerably reduced by p-type doping, the reverse transition of conduction type has not yet been reported [17]. Even though holes are generated, a larger number of electrons are also generated from oxygen defects, i.e., the self-compensation effect occurs. Although Cu_2O is also expected to serve as a bipolar semiconductor because of its carrier stability, the reverse transition from p-type to n-type has not yet been reported.

Here we need to notice the above argument predicts the tendency of carrier dupability and more specific consideration is required for some cases. An example is CdO for which both VBM and CBM are located in the doping window in Figure 2.7. However, CdO exhibits n-type conduction, but no p-type conduction has been reported notwithstanding various efforts. The difficulty in p-type doping is due to easier formation of oxygen vacancy leading to electron-doping i.e., hole-killer defect.

Figure 2.8 shows the relation between various defect formation energies in CdO and Fermi level [17]. It is evident from the figure that the formation energy of neutral vacancy (trapping 2 electrons) is lower than that of Cd interstitial (Cd_i) leading to hole doping in the whole bang gap region and under Cd-rich/poor conditions. This example illustrates whether really carrier doping occurs or not, depending on the easiness of counter defect formation even the band edge of the semiconductor locates within the doping window [18].

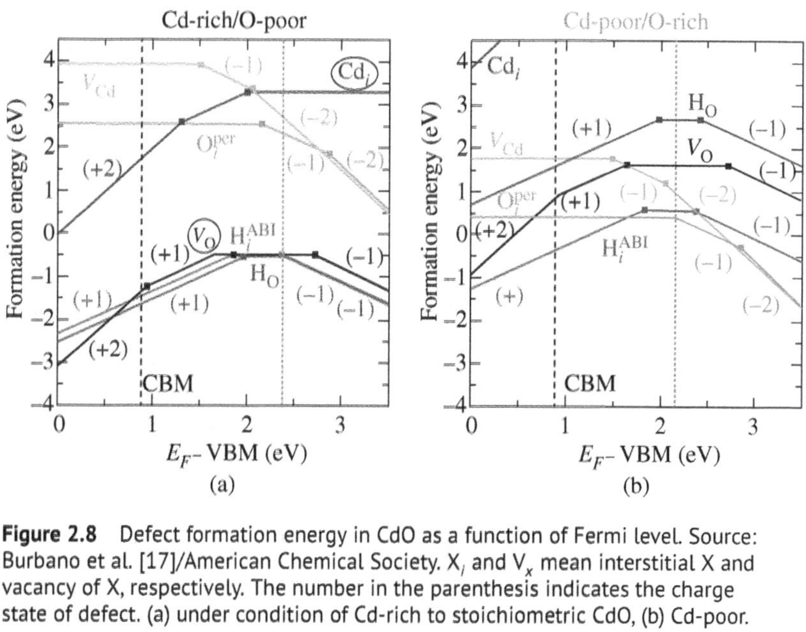

Figure 2.8 Defect formation energy in CdO as a function of Fermi level. Source: Burbano et al. [17]/American Chemical Society. X_i and V_x mean interstitial X and vacancy of X, respectively. The number in the parenthesis indicates the charge state of defect. (a) under condition of Cd-rich to stoichiometric CdO, (b) Cd-poor.

2.3 Amorphous Oxide Semiconductors for TFT Applications

2.3.1 Amorphous Semiconductors

While amorphous hydrogenated Si (a-Si:H) had been exclusively used as the backplane TFTs to drive flat panel displays, amorphous oxide semiconductor (AOS) TFTs are becoming a strong alternative due to their high mobility, low off-current, high transparency, and easy fabrication. In particular, AOSs with In–Ga–Zn–O compositions (IGZO) have practically been applied in the backplanes for high-resolution and energy-saving LCDs and large-sized OLED TVs.

Condensed matter may be classified into two categories, crystalline and amorphous materials, depending on whether unit cell exists or not. This discrimination is usually performed by X-ray diffraction. Amorphous materials have distinct advantages over crystalline materials, i.e., large-sized and homogeneous (grain-boundary-free) thin films can be fabricated at low temperatures. In addition, material properties may be tuned by continuously varying the chemical composition. The essence of semiconductors is the controllability of Fermi Level (E_F) by intentional operation such as impurity doping and biasing. If the advantage of amorphous materials and the essence of semiconductors could be merged, the resulting amorphous semiconductor should be an ideal semiconductor for giant microelectronics represented by flat panel displays as illustrated in Figure 2.9.

However, amorphous semiconductors reported so far are far from this ideal because of very high defects (chemical disorder in local structure like vacancy) concentration and tail state (localized states induced by disorder in medium range structure) density [19]. Figure 2.10 shows a schematic drawing of electronic states of amorphous semiconductors. Since these defects and tail states work as charge trapping sites, the control of Fermi level in amorphous semiconductors is generally quite difficult. Thus,

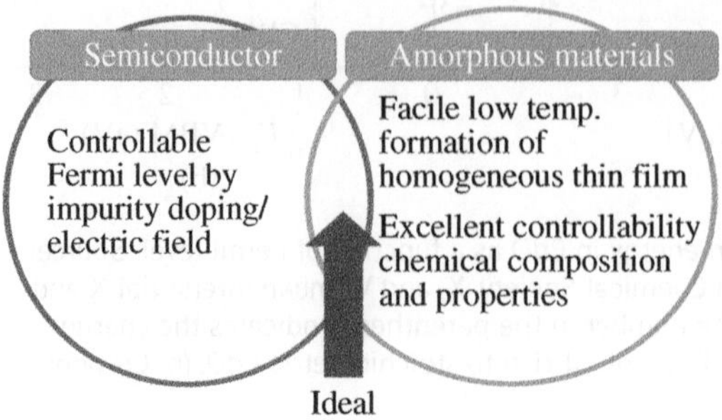

Figure 2.9 Ideal amorphous semiconductor meeting both advantages of semiconductor and amorphous materials.

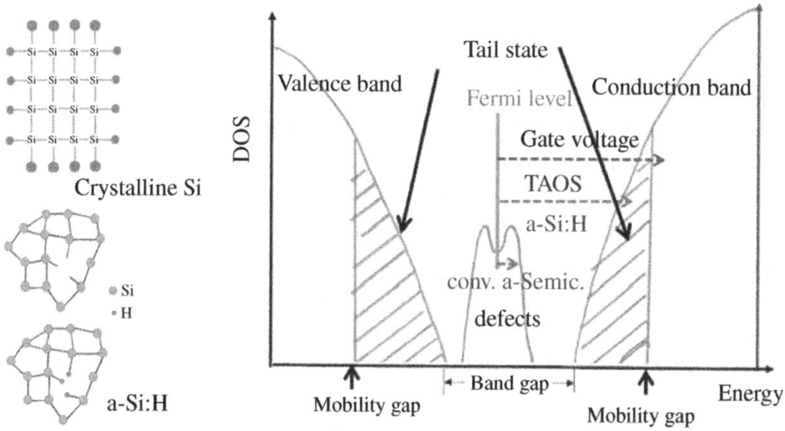

Figure 2.10 Schematic energy state of amorphous semiconductors. Fermi level controllability: TAOS > a-Si:H ≫ conventional amorphous semiconductors including H-free a-Si.

charge transport occurs by hopping among localized states as reported for amorphous chalcogenides and semiconducting oxide glasses based on V_2O_5. The requirements for semiconductor TFT channels are much more severe than a simple semiconducting nature. Although conventional amorphous semiconductors do not work as TFT channel materials due to high concentrations of localized state. A-Si:H is a well-known first amorphous semiconductor in which E_F is controllable by biasing/impurity doping [20, 21]. This is the primary reason why a-Si:H attracted much attention as materials for large-area electronics in which crystalline Si is difficult to deal with. The large reduction of Si dangling bonds responsible for mid-gap levels by passivation with hydrogen makes it possible to shift E_F to band edges, but E_F cannot be beyond the mobility edges due to high tail state density arising from the medium range disorder. Thus, the mobility of a-Si:H TFTs remains 0.5–1 $cm^2/V\,s$, which is lower by 2 orders of magnitude than that of polycrystalline Si-TFTs.

TAOS based on post-transition metal oxides is the first category of amorphous semiconductors in which E_F is controllable beyond the mobility edge in the conduction band leading to the realization of band conduction [22]. Thus, TAOS TFTs exhibit large mobility, >10 $cm^2/V\,s$, which is comparable to that in the corresponding polycrystalline thin films. Another unique advantage of TAOS TFT over Si-TFTs is its low off-current leading to low power consumption because the duty ratios of TFTs in displays and memories are very small such as <0.01 [23]. This chapter describes the origin of such favorable properties of TAOS TFTs in comparison with Si-TFTs.

2.3.2 Transparent Amorphous Oxide Semiconductors

2.3.2.1 Materials Design

What happens if these transparent oxide semiconductors described previously become amorphous state? In amorphous materials, structural disorder concentrates on energetically weak structural units. In most cases, the structural disorder appears prominently as a broad distribution of bond angle. When the bond angle has a large distribution, how is the effective mass (in other words, the transfer rate between neighboring atom's s orbitals or overlap integrals) modified and affect the transport of carrier electrons? We considered the two cases: (i) covalent semiconductors and (ii) ionic semiconductors. In the former case, the magnitude of the overlap between the unoccupied sp-hybridized orbitals with high spatial anisotropy is very sensitive to the variation in the bond angle. As a consequence, rather deep localized states would be created at somewhat high concentrations; thereby, the drift mobility would be largely reduced. On the other hand, the magnitude of the overlap in the latter case differs depending on the choice of metal cations; when the spatial spread of metal cation's *s* orbital, corresponding to the lowest unoccupied atomic orbital, is larger than the inter-cation separation, the magnitude should be insensitive to the bond angle distribution. The reasons are two; the first is the orbital radius of vacant s-orbital distinctly larger than the ionic radius which corresponds to the size of the occupied outermost orbital. The second is that *s* orbitals are isotropic in shape, unlike p-orbital. As a consequence, we may anticipate that these ionic amorphous materials have large electron mobility comparable to that in the corresponding crystalline phase. In the case that the spatial spread of the metal *s* orbital is small, such a favorable situation cannot be expected [22, 24]. The spatial spread of the *s* orbital of a metal cation is primarily determined by the principal quantum number (n) and is modified by the charge state of the cation as discussed for the crystalline TOSs in Chapter 1. Thus, candidates for high mobility TAOS are found in oxides of post-transition metal (PTM) cations with an electronic configuration $(n-1)d^{10}ns^0$, where $n \geq 5$ [22]. Note that in crystalline oxide semiconductors, this requirement is relaxed to $n \geq 4$ as exemplified by ZnO and Ga_2O_3 with the $(3d)^{10}(4s)^0$ configuration. Figure 2.11 draws the difference in orbital constitution at CBM between Si and a PTM oxide and between crystalline and amorphous phases [25]. The drastic reduction of the electron mobility in the amorphous state from c-Si may be understood in terms of distinct reduction of the overlap of the sp^3 orbitals, whereas the mobility in c-PTM oxides is almost reserved even in the amorphous state.

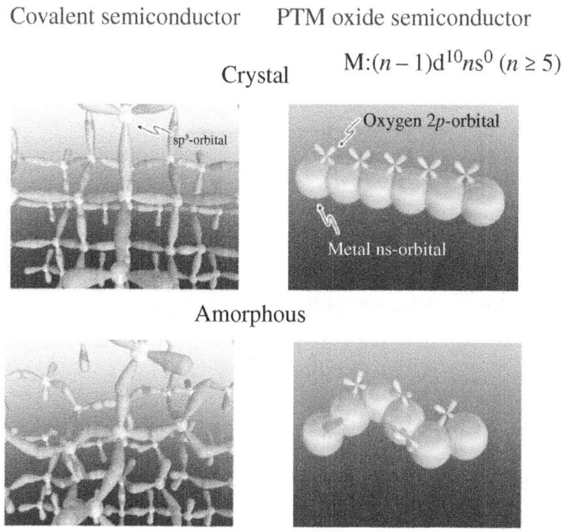

Figure 2.11 Orbital drawing of conduction minimum in covalent semiconductors like Si and post-transition metal oxides with largely spread metal s-orbital. Source: Nomura et al. [25]/ with permission of Springer Nature.

In a sense, the situation of CBM in PTM oxides is similar to that in amorphous metal alloys (conductivity is slightly lower than that in crystalline alloy) [26] in the aspect that metal orbitals dominantly constitute the electron pathways, as shown in Figure 2.12. This idea was examined for $2CdO{\cdot}GeO_2$ thin film for which Cd^{2+} with $4d^{10}5s^0$ configuration meets the above criterion for TAOS.

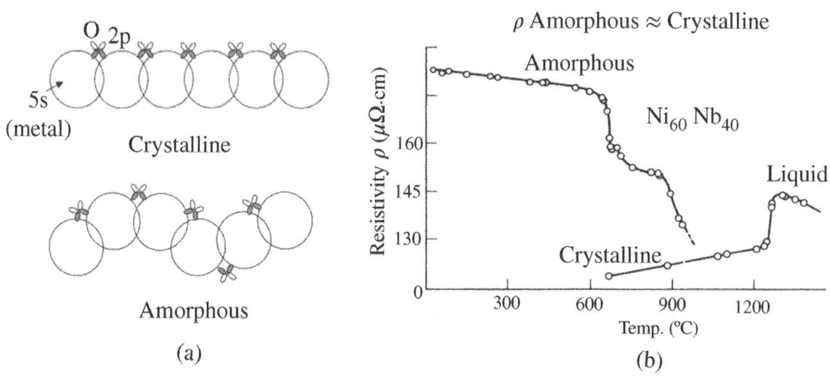

Figure 2.12 Analogy between TAOS and amorphous metals. (a) Schematic drawing of metal s-orbital in TAOS. Since this s-orbital is like a metal atom in amorphous metals, the situation is similar. (b) Electrical resistivity of crystalline and amorphous (solid and molten) metals. Please note the resistivity is expressed in linear scale, not logarithmic scale.

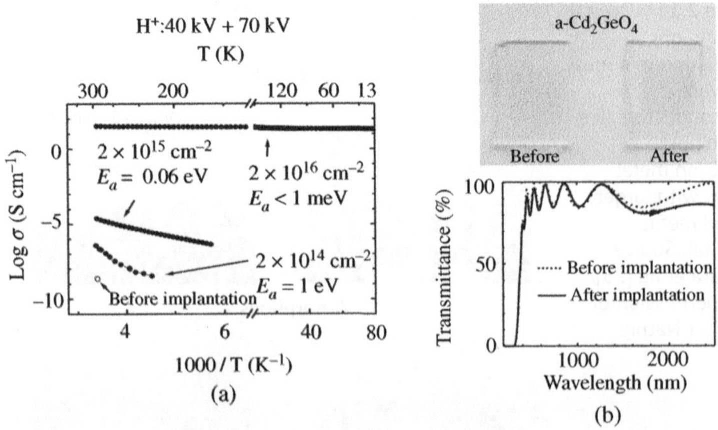

Figure 2.13 Amorphous Cd_2GeO_4 thin films homogeneously implanted with proton at 40 keV + 70 kV. (a) Electrical conductivity, (b) optical absorption.

Experimental results, as shown in Figure 2.13, on this thin film show that the Fermi level is continuously controllable from the mid-gap to CBM by proton implantation. The Hall mobility (mobility of electron in the conduction band) of amorphous thin films (~10 cm^2/V s) is close to that in the polycrystalline thin films [27], which is as predicted by the idea described above.

The validity of this idea is demonstrated quantitatively by a strategy summarized in Figure 2.14. The density of states (DOS) of conduction band can be measured by inverse photoelectron spectroscopy. The DOS is computed based on atomic positions determined by analysis of the measured X-ray radial distribution function with the reverse Monte Carlo simulation, which is then compared with the measured DOS.

Figure 2.15 illustrates the connectivity of Cd 5s orbitals at the CBM of amorphous 2CdO-GeO_2. Here, two Cd^{2+} ions were connected by a line for visualization of orbital overlap when 2 × Cd 5s Slater orbital radius is larger than the interatomic separation. It is clearly observed the lines are 3D-connected throughout the sample forming a percolated electron-path way [28].

Due to the above requirement for high-mobility oxides, major TAOS materials need to contain In^{3+} or Sn^{4+}, each of which has $(4d)^{10}(5s)^0$ configuration, and its fraction is required to be beyond the percolation threshold of its 5s orbitals at the CBM. There is a distinct difference between In and Sn in oxide materials. The valence state of In is stable as +3, while Sn takes two stable charge states, +4 and +2, depending on its environment. Since the filled

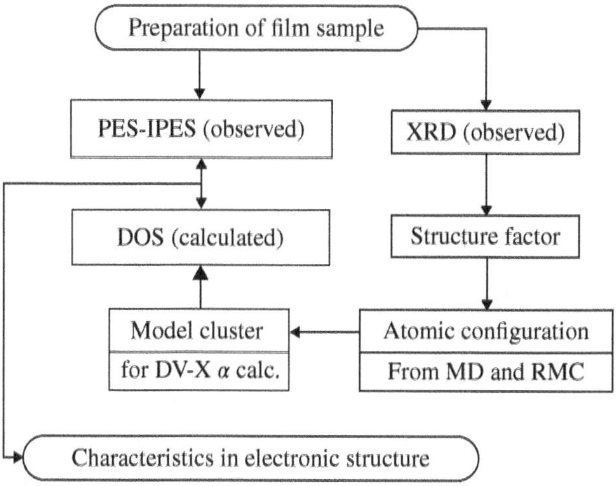

Figure 2.14 Approach to elucidation of CBM electronic state in TAOS thin film by combined experimental and computational methods.

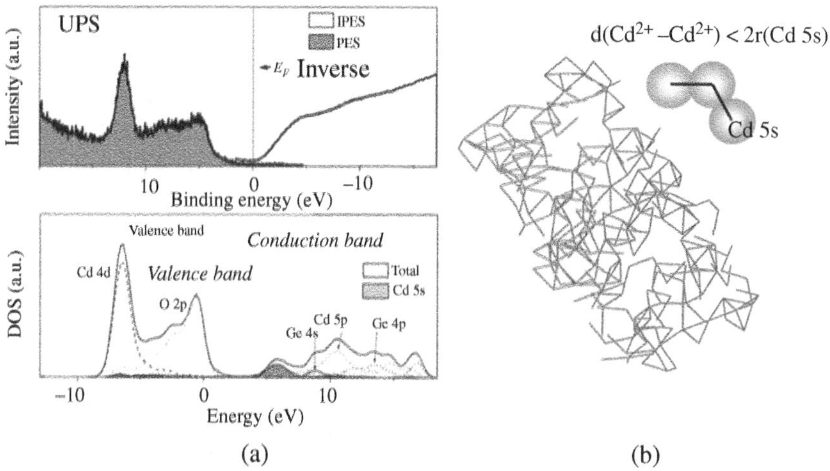

Figure 2.15 Electronic Structure of a-Cd_2GeO_4. (a) Density of states, top: observed by photoemission (UPS for valence band, Inverse PE for conduction band), bottom; calculated using atomic coordinates determined by inverse Monte Carlo analysis of measured radial distribution function, (b) connectivity of Cd 5s-orbitals at the conduction band minimum. When twice of Cd5s orbital radius is larger than the Cd-Cd separation, the two Cd ions are connected by a line to visualize the orbital connectivity.

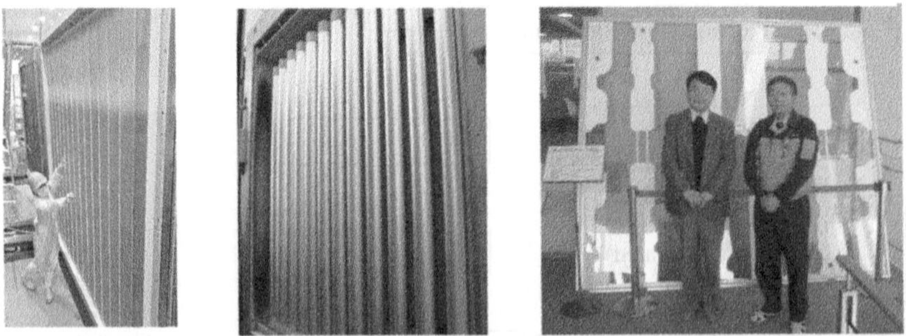

(a) (b) (c)

Figure 2.16 Photos of IGZO ceramic target for sputtering: (a) strip-type, (b) rotary-type, and (c) IGZO thin film deposited on G8.5-sized glass plate.

5*s* states are located at above VBM, Sn^{2+} with $5s^2$ configuration does not work unlike Sn^{4+} for the electron conduction path formation. TAOS materials with many different compositions have been reported to date and the materials which work as excellent TFT channels are almost restricted to In- and/or Sn-containing systems.

Here, a-In–Ga–Zn–O (a-IGZO) [25, 29] is taken as a representative TAOS material. $InGaZnO_4$ is a stable compound in the phase diagram and the synthesis of such a ceramic target for sputtering is facile just by heating the mixture of In_2O_3, Ga_2O_3, and ZnO at high temperatures in an ambient atmosphere [28, 29]. This is a factor why IGZO was adapted as the TFT channel for displays. Figure 2.16 is the photo of IGZO ceramic targets for the deposition of a-IGZO thin films on large-sized glass substrates.

Figure 2.17 shows the Hall mobility and electron carrier concentrations in amorphous IGZO thin films deposited at RT on glass substrates. Amorphous thin films are obtained all over the region except the vicinity of pure ZnO and In_2O_3 [25]. Both the mobility and carrier concentrations increase with In_2O_3 content and the maximum values in the amorphous thin films are seen around 30% ZnO–70% In_2O_3, which corresponds to that of commercial amorphous TCO. It is obvious from the figure that the In^{3+} ions contribute to a large electron mobility and are thus called a "*Mobility Enhancer*." Ga with larger ionic strength (ionic charge/ionic radius) forms stronger chemical bond with oxygen and would suppress the formation of oxygen vacancy leading to the generation of conduction electrons, and thus called "*Carrier generation Suppressor*" [26]. The role of the suppressor is critical for TFT application because the conductivity is controlled by the gate voltage, which is almost impossible for films having the high carrier concentration. Although a variety of chemical compositions are possible

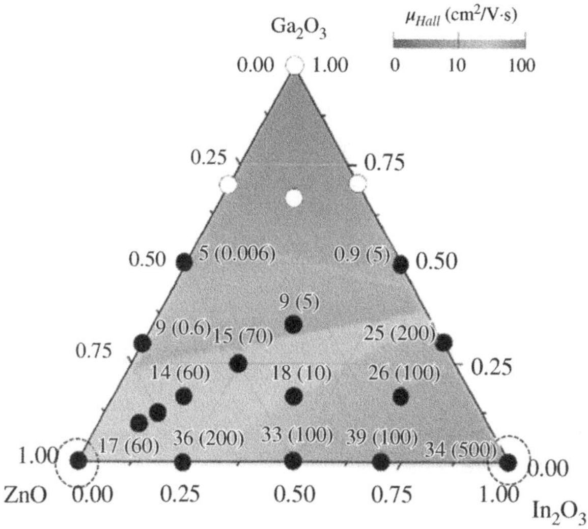

Figure 2.17 Hall mobility and electron carrier concentrations ($\times 10^{18}$ cm^{-3}) in amorphous In–Ga–Zn–O thin films. Samples encircled by dotted line (pure In_2O_3 and ZnO) are not amorphous but polycrystalline.

for TAOS materials, these mobility enhancers and carrier generation suppressors are essential components in TAOS channel materials for high-performance TFTs. Zinc appears not to be essential but is expected to stabilize amorphous structure taking tetrahedral coordination like "*network former*" in glass. For practical application, a large-sized, dense ceramic target for sputtering is necessary for the deposition of large-area thin film on glass plates.

2.3.2.2 Characteristic Carrier Transport Properties

TAOS has several common and unique properties which are not seen in conventional amorphous semiconductors:

(1) Large electron mobilities >10 cm^2/V s, which is higher by 1–2 orders of magnitude than that in a-Si:H.
(2) A degenerate state can be realized. This is different from the other amorphous semiconductors.

c-Si is easily changed to the degenerate state by impurity doping (~10^{16} cm^{-3}), but no such state is attained in a-Si:H. On the other hand, in TAOS, the Fermi level (E_F) can exceed the mobility gap easily by carrier doping, leading to band conduction. It is considered that this striking difference originates from the chemical bonding nature between the materials,

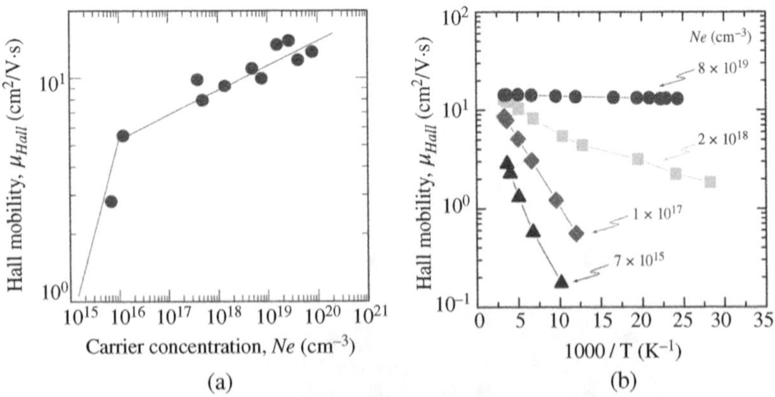

Figure 2.18 Hall mobility in a-IGZO thin films as a function of (a) carrier electron concentration and (b) temperature.

i.e., strong ionic bonding with spherical potential (r^{-1}; r = interatomic separation) is much favorable to form a shallow and small density of tail states compared with covalent bonding with steep and anisotropic potential).

Hall effect measurements, which is a standard method to evaluate the carrier concentration and mobility in crystalline semiconductors, cannot be used for conventional amorphous semiconductors [19]. The reason is that the mean free path of the carrier is so short, e.g., comparable or smaller than inter-atomic separation. On the contrary, TAOS materials give distinct Hall voltages and evaluated N_e and mobility of carriers are reliable because the electron mean free path is several nanometers, which is distinctly larger than the interatomic separation. A typical carrier transport property of a-IGZO is shown in Figure 2.18 as an example of TAOS. It is noted that Hall mobility (μ_{Hall}) largely depends on carrier concentration (N_e) due to the presence of potential barriers arising from structural disorder. μ_{Hall} increases with increasing N_e and finally exceeds 10 cm^2/V s if N_e exceeds ~10^{18} cm^{-3}. The activation energy of mobility is continuously decreased with N_e and eventually degenerate state is realized at $N_e = 10^{19}$~10^{20} cm^{-3}.

This μ_{Hall} vs. Ne behavior is explained by the percolation conduction model, in which electron transport is controlled by distributed potential barriers above CBM, as shown in Figure 2.19 [30, 31]. If the TAOS has a high concentration defect or with very low N_e, E_F is below the percolation threshold level (E_{th}), such as at very low temperatures, hopping conduction is dominant like conventional amorphous semiconductors [32].

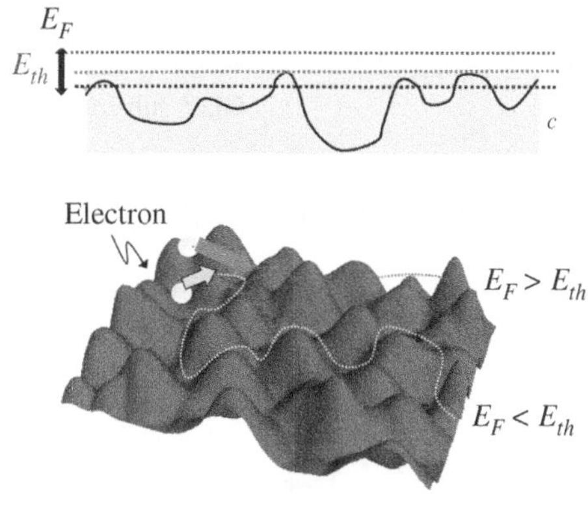

Figure 2.19 Schematic illustration of near the conduction band bottom of TAOS. Arrows denote conduction paths of electron for the cases of $E_F > E_{th}$ or $E_F < E_{th}$.

2.3.3 Amorphous Oxide Semiconductor TFTs

Figure 2.20 draws the typical fabrication procedure of TAOS TFTs with the bottom-gate structure. Detailed discussions on fabrication of oxide TFTs with various structures are included in Chapter 4.

Figure 2.21 shows representative characteristics of (a) an a-IGZO (In:Ga:Zn = 1:1:1) TFT (b) an a-Si:H TFT both fabricated on glass substrates. The on-current (I_{on}) of the former is 2 orders of magnitude larger than that of the latter. In general, an OLED pixel requires several μA of driving current, indicating that only 5 V is enough for an a-IGZO TFT which is not achievable for an a-Si:H TFT with comparable size. This is the reason why the a-Si:H backplane is not used in OLED displays. Only oxide or LTPS TFT backplanes are available for this kind of display. The a-IGZO TFT exhibits a high μ_{FE} of >10 cm²/V s and a low I_{off} of ~10^{-14} A near the conventional measurement limit, which is distinctly lower than that of an a-Si:H TFT. In addition, the low operation voltage a-IGZO TFT benefits from the small subthreshold voltage (V_{th}) and subthreshold slope (SS).

To obtain good TFT performance, the choice of the sample having appropriate conductivity is important. Figure 2.22 shows the conductivity in a-IGZO thin films as a function of P_{O2} during film deposition. One may see that good TFTs are obtained with a-IGZO channels having the electrical conductivities $10^{-6}-10^{-3}$ S cm^{-1}, corresponding to $N_e = 10^{12} - 10^{15}$ cm^{-3} assuming $\mu_{Hall} = 10$ cm²/V s.

A key procedure for obtaining practical TFTs is post-deposition thermal annealing at appropriate conditions. Although TAOS TFTs operate with large $\mu_{FE} > 10$ cm²/V s even without substrate heating or thermal annealing as shown in Figure 2.23a; however, unannealed TFTs exhibit serious hysteresis and instability, e.g., a large V_{th} shift (ΔV_{th}), as shown in Figure 2.23b.

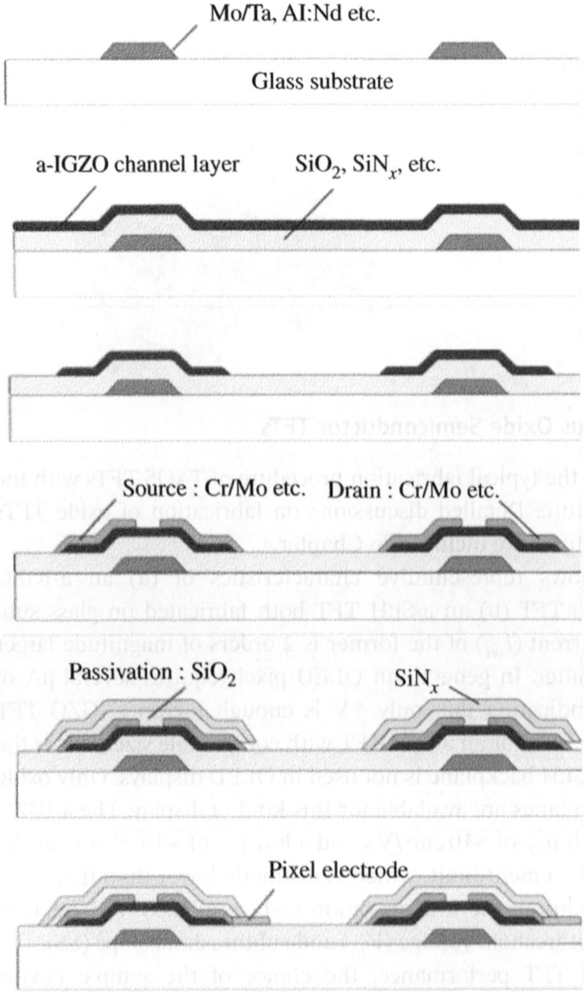

Figure 2.20 Fabrication process of a bottom-gate structured TAOS TFT.

Typically, clockwise hysteresis is observed (both for unannealed and annealed TFTs), suggesting effects of shallow electron traps in the channel. For typical instability modes, positive ΔV_{th} are observed for positive V_{GS} bias tests but negative ΔV_{th} for negative V_{GS} (usually the negative ΔV_{th} is small or even zero). Post-deposition thermal annealing at 300–400 °C in air or O_2 improves these issues [33]. Thermal annealing is also effective in homogenizing TFT characteristics and improving uniformity, as shown in Figure 2.23c.

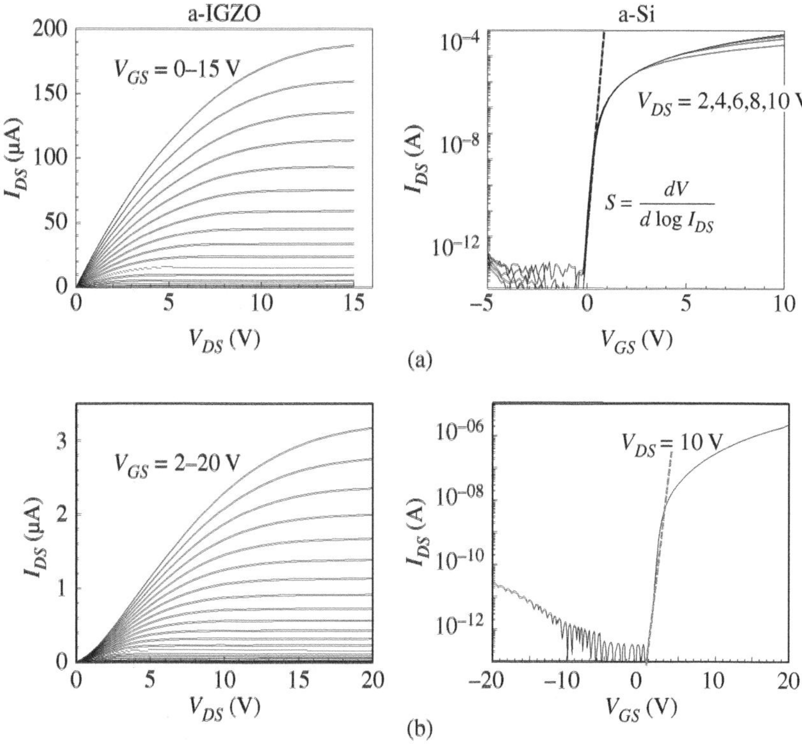

Figure 2.21 Typical TFT characteristics of (a) a-IGZO and (b) a-Si TFTs. (Left) Output and (right) transfer curves.

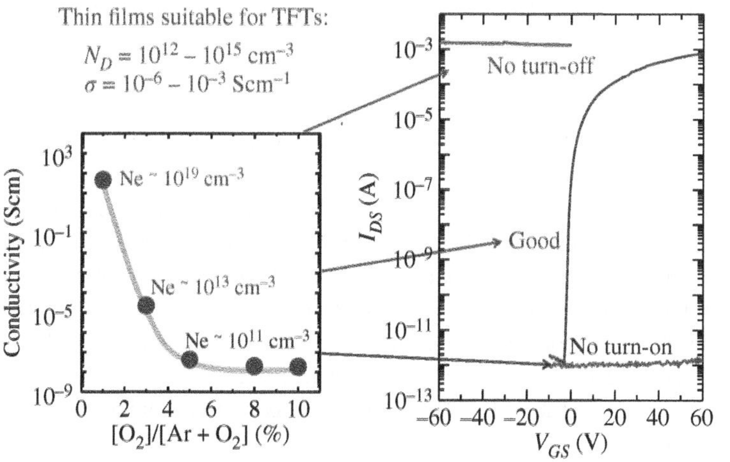

Figure 2.22 Transport properties in a-IGZO thin films suitable for TFTs.

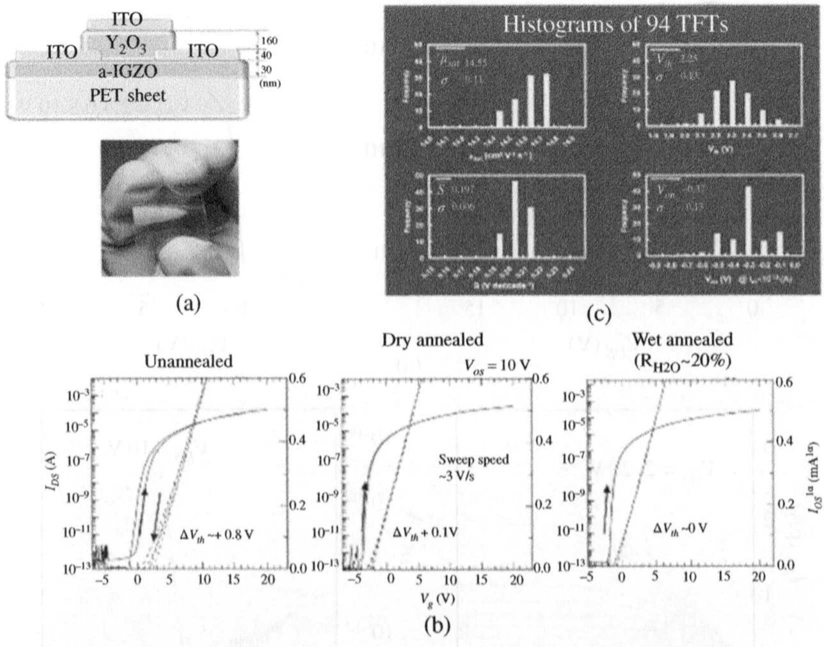

Figure 2.23 IGZO TFTs (a) on a flexible substrate, (b) effects of 350 °C post-annealing on transfer characteristics, (c) performance histograms of ~100 TFTs fabricated on a glass substrate and post-annealed at 350 °C in air.

Here, the reason why there is such a large difference in mobilities between a-IGZO and a-Si:H TFT is considered. The tail state density and mid-gap trap density evaluated from the TFT characteristics assuming exponential tail state density and Gaussian-shape mid-gap trap [34], are shown in Figure 2.24. It is evident that the tail state density between the CBM and E_F is higher by 2 orders of magnitude than that of a-IGZO and no large distinct peak of the mid-gap trap state is seen for a-IGZO unlike a-Si:H. Such a lower tail state density and the mid-gap density in a-IGZO make it easier to upshift E_F beyond the mobility gap (>CBM) inducing band conduction. This striking difference in tail state and mid-gap state densities originates from chemical bonding and coordination; a-Si:H is composed of covalent Si-Si and Si—H bonds and each Si takes tetrahedral coordination. Thus, the structure has a network-like and a large distribution in Si—Si—Si (H) bond angles, which gives rise to large variation in Si—Si bonding/anti-boning levels constituting the VBM/CBM. As a consequence, a large tail state density is created.

When an n-type dopant, such as P, is added into a-Si:H, it substitutes the Si site, but the generated electron is trapped on Si to form a lone pair which

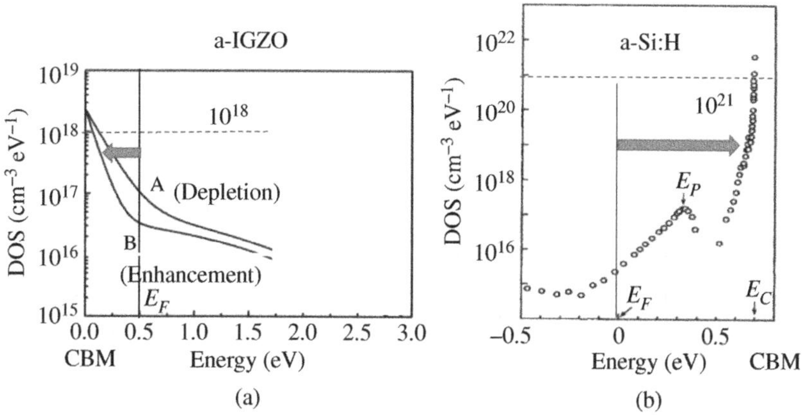

Figure 2.24 Density of state below conduction band minimum of (a) a-IGZO and (b) a-Si:H. Data on a-IGZO were evaluated from analysis of transfer curves of TFT. Source: Adapted from Kimura et al. [34].

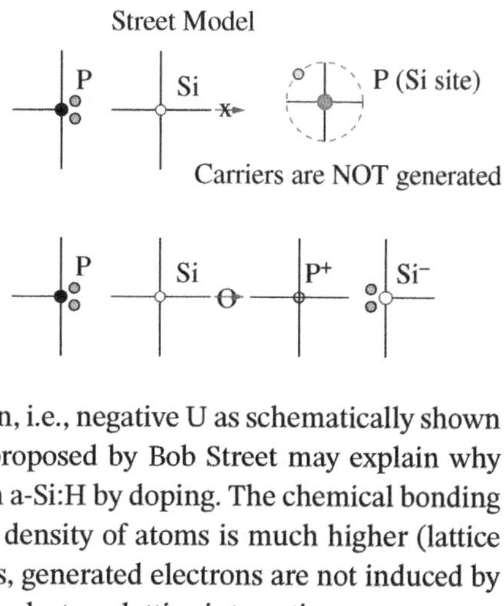

Figure 2.25 Model for why Fermi level cannot exceed the mobility gap in a-Si:H. Electron generated by P-substitution to Si is used to create a lone pair electron of 3-coordinated Si. Source: Adapted from Street [35].

is stabilized by the lattice distortion, i.e., negative U as schematically shown in Figure 2.25 [35]. This model proposed by Bob Street may explain why band conduction does not occur in a-Si:H by doping. The chemical bonding in TAOS is ionic and the packing density of atoms is much higher (lattice is much harder) than a-Si-H. Thus, generated electrons are not induced by negative U originating from strong electron-lattice interaction.

2.3.4 Electronic States and TFT Operation

TFTs are devices that source-drain current is modulated by applying a voltage to gate insulator over several to ten orders of magnitude. Thus, the density of the mid-gap state and tail state in channel semiconductors are critical for TFT operation because these states work as carrier traps. The dominant factor of μ_{FE} is partly different from that of μ_{Hall} because μ_{FE} is expressed roughly by $\mu_{Hall}\,(N_{ind} - N_{trap})/N_{ind}$, where N_{ind} is the total electron density

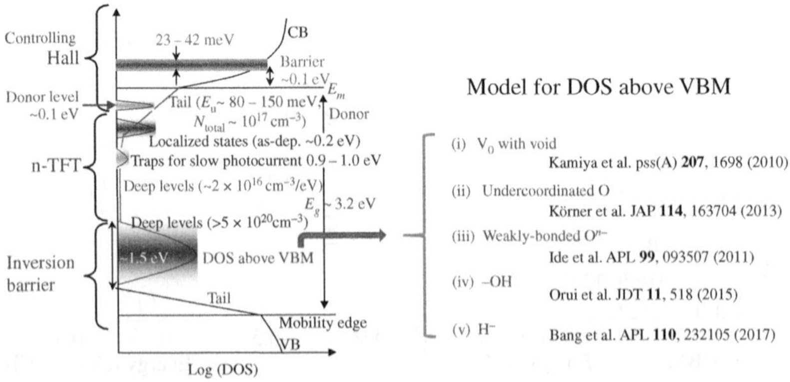

Figure 2.26 Electronic states of a-IGZO and relevant TFT operation conditions. Right is a list of models for large DOS above VBM.

induced by the gate voltage and N_{trap} is the density of the induced electrons trapped by mid-gap defects and tail states. Therefore, low N_{trap} is important to obtain high μ_{FE}, and the low N_{trap} is confirmed by TFT analyses, as shown in Figure 2.26, *C-V* analyses, etc. These studies have revealed that these localized state densities in TAOS are 2–3 orders of magnitude smaller than that of a-Si:H where E_F is close to CBM. This striking difference is the main reason why the band conduction can be induced in TAOS but not in a-Si:H by biasing gate voltage as described in Section 2.3.3.

Figure 2.26 illustrates the electronic structures of a-IGZO revealed by the experimental researches [36, 37]. Noteworthy is the presence of a large DOS (10^{19-20}cm^{-3}) above the CBM. This DOS was first observed [38] for the thin films by Hard X-ray photoemission spectroscopy (HAX-PES) using a synchrotron radiation facility (6–8 keV). HAX-PES has two advantages over conventional lab-PES; the first is to be bulk sensitive originating from a large escape depth (several nanometers from the top surface) of a photoelectron [39]. This makes it possible to get reliable information from as-prepared thin films without any surface treatment to prepare the fresh surface such as sputtering. The other is a large ionization cross-section even for *s*-state electrons which are associated with oxygen vacancy trapping electrons, hydrogen anion, and low valence metal cations with ns^2 electron configuration such as In^+ and Ga^+. Since this DOS in question is located above VBM and far below E_F, n-channel TFT operation does not suffer directly from their influence, fortunately. However, this DOS plays a crucial role in energy saving of the devices and TFT degradation under the display dominant operation mode of off-state (negative gate biasing under backlight illumination stress) which is called NBIS instability. The E_F cannot push down under negative gate voltage due to the presence of this large DOS, i.e., the p-channel does not open when a negative bias is applied (note that p-channel is not open in TAOS TFTs if this DOS is not, because no p-type conduction does not occur due to deep VBM and large hole mass as described in Chapter 1 and Section 2.2.3). This feature makes a contrast with Si-based TFTs such as low-temperature processed polysilicon (LTPS) and a-Si:H TFT which shows ambipolar operation, i.e., the drain current is increased when a negative bias is applied to the gate. Since TFTs for LCDs dominantly stay off-state, the unipolarity of TAOS TFTs leads to energy saving. This is the reason why the off current of IGZO-TFTs is much lower than a-Si:H and LTPS-TFTs, resulting in low power consumption. This large DOS above VBM is observed for ZnO and crystalline IGZO and would be considered common in oxide semiconductors.

This large DOS above VBM works as the source of NBIS instability which is induced by subgap light illumination under negative gate bias. That is, the electron in the DOS is excited to CBM by subgap illumination (~2.3 eV). The electron excited to CBM diffuses away due to the high electron mobility at CBM, while the remaining hole drifts to the interface between the channel and the gate insulator and is trapped there to form a fixed positive charge [40]. As a result, the threshold voltage shifts to the negative voltage side (NBIS instability) was first reported in poly-crystalline ZnO-TFTs [41] and was found in almost all the n-type oxide TFTs irrespective of crystalline or amorphous. This suppression is an important issue for the practical application of oxide-TFTs for displays because the threshold voltage shifts result in output current changes, in particular for OLEDs which are driven by current. In practical display applications, this NBIS instability is effectively reduced by adopting the top-gate, top-contact device structure (contact metal shields the backlight). There are several models for the origins of this large DOS as described in Figure 2.26. Although each model is more or less associated with oxygen-deficient-type structural imperfection and appears to work as acceptors, no evidence has yet been reported to conclude which model is the best at the present stage. Multiple defects appear likely to coexist depending on the fabrication conditions.

2.3.5 Features of Transparent Amorphous Oxide TFTs

TAOS TFTs have several unique features in comparison with other representative TFTs [22, 42]. Figure 2.27 shows schematic TFT characteristics of a-Si:H, LTPS and a-IGZO along with a detailed comparison including industrial production, as shown in Table 2.1. AOS TFTs are now widely applied to the backplane to drive pixels of high-resolution LCDs and

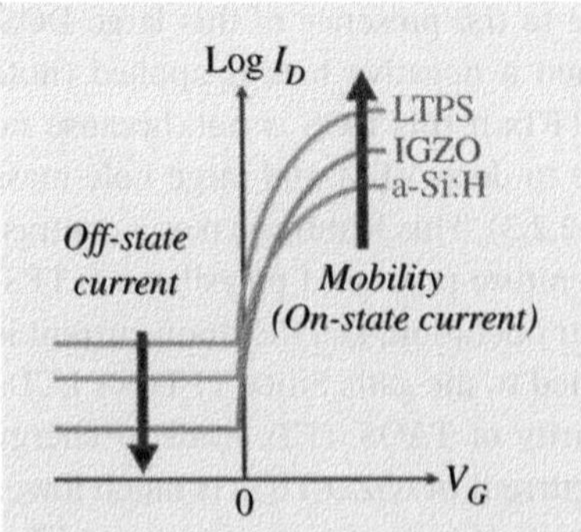

Figure 2.27 Transfer characteristics of three representative TFTs.

Table 2.1 Comparison in performance of representative TFTs.

Property	a-Si:H	LTPS	IGZO
Structure	Amorphous	Polycrystalline	Amorphous
Mobility (cm²/V s)	~1	50~100	10~30
Off current	Fair	High	Low
VT stability	Poor	Good	Fair
Fabrication process	Simple (plasma CVD)	Complex (laser crystallization)	Simple (sputtering)
Scalability	Good	Poor	Good
C-MOS	X	O	X
Optical transparency	Low	Low	High
Starting materials	SiH_4 gas (dangerous)	a-Si:H	Oxide ceramics

OLED-TV utilizing the features i, ii, and iii. Five characteristics are raised for a-IGZO over a-Si:H and LTPT.

i. Mobility is 10–30 cm²/V s depending on the In content. These values are higher by an order of magnitude than a-Si:H but are lower by 1/5~1/3 than LTPS.

ii. No inversion (p-channel) occurs when the gate voltage is reversed [36, 43].

iii. Large-scale production is possible because the thin film deposition is made at RT by a conventional sputtering technique using a stable ceramic target.

iv. TFT performance and stability are sensitive to the partial O_2 pressure during deposition and post-annealing conditions [44]. Structural-relaxation-driven electron doping of amorphous oxide semiconductors by increasing the concentration of oxygen vacancies in shallow-donor states [45].

v. Optically transparent. This feature is favorable to enhance the aperture ratio leading to transparent displays.

vi. Possible to prepare using the solution process. Since oxide is chemically and thermally stable in the ambient atmosphere, oxide-TFTs may be fabricated via solution processing and subsequent post-annealing.

2.4 Fundamental Material Issues in High-performance Oxide TFTs

Oxide TFTs, such as IGZO-TFTs, have a wide range of applications in displays, sensors, and circuits, which are shown in Chapter 5. For many advanced applications, the oxide semiconductor material properties are critical to the TFT performance. Figure 2.28 shows critical issues in the high-resolution displays, logic circuits, and memories [46]. The following are discussions on three important topics affecting the performance of the oxide TFT.

2.4.1 CO Effect on Mobility and Stability of Oxide TFTs

TFT mobility and stability are in trade-off relation in many cases. A typical example is a-IZO (ZnO content 10%) material. The amorphous structure of

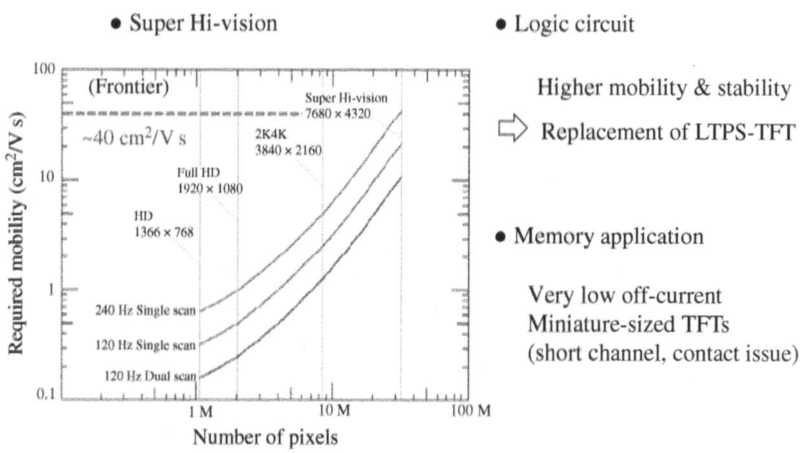

Figure 2.28 Relation between TFT mobility and number of pixels for different display performance [46] and new demands. Source: [46]/with permission of Yojiro Matsueda.

In_2O_3 is stabilized by incorporation of ZnO but the Hall mobility (μ_{Hall}) of In_2O_3 does not degrade. Conventionally sputtered thin films of this TAOS material are used as amorphous TCO (transparent conductive oxide, i.e., metallic conducting), not semiconductors. When IZO is sputter-deposited in highly oxidizing conditions such as high partial O_2 pressure, the resulting thin films with low N_e enough to use as a TFT channel layer can be fabricated. TFTs based on such a thin film exhibit high μ_{FE} such as 50 cm^2/V s is obtained but the off-current is continuously increased with time and eventually lose device performance.

Higher mobility AOS TFTs are obtained in In_2O_3-SnO_2-ZnO (ITZO) system. Replacement of Ga^{3+} with Sn^{4+} enhances the mobility because Ga^{3+} does not form the conduction band bottom but Sn^{4+} with the same electronic configuration as In^{3+} can do, and Sn^{4+} with high ionic potential Z/R is expected to serve as suppressor like Ga^{3+}. However, the instability of threshold voltage under gate-biasing is much worse in ITZO than IGZO. This is a commonly observed in oxide TFTs and is known as trade-off between mobility and stability.

Recently, high mobility-high stability AOS-TFTs were reported through the understanding of the underlined mechanism. CO-related impurities originating from a trace amount of residual photoresist used in the lithographic process were found to be a major origin of negative bias instability [47]. Eliminating CO-related impurities, combined with subsequent surface passivation, allows to fabricate stable ITZO TFTs with a mobility of 70 cm^2/V s . Figure 2.29 shows transfer curves of obtained high mobility ITZO-TFTs having good stability for NBS(a), PBS(b) and NBIS(c). Detailed discussion on bias-stress instability and negative bias illumination stability are discussed in Chapter 3.

Carbon monoxide CO is adsorbed on the top surface of TAOS thin films trapping an electron, as shown in Figure 2.30a [48]. When the negative gate voltage is applied, the adsorbed CO releases an electron, which is injected into the CBM inducing the negative ΔV_{th}. The sensitivity of ΔV_{th} to CO adsorption depends on mobility in TAOS materials. ITZO-TFTs with higher mobility exhibit distinctly larger ΔV_{th} than IGZO TFTs with lower mobility when the amount of adsorbed CO is similar. This sensitivity difference is attributed to the energy difference between CBM and donor level corresponding to the activation energy (E_a) of the donor for band conduction as schematically drawn in Figure 2.30b. That is, the high mobility system (ITZO) has a lower CBM than the low mobility system (IGZO). Here, donor levels are assumed to be almost constant in TAOS materials. Then, the carrier concentration in the high mobility system becomes higher than that in the low mobility system (IGZO) [48]. For example, E_a values in typical

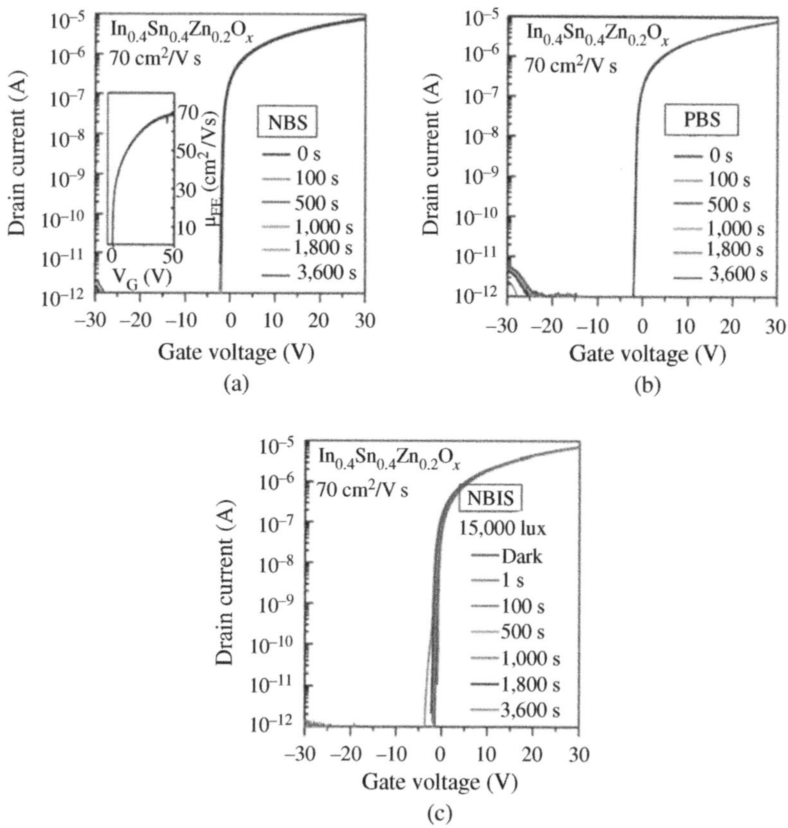

Figure 2.29 High mobility and high stability a-ITZO TFTs. (a) negative gate-bias sress (NBS), (b) positive gate-bias stress (NBS), and (c) negative gate-bias illumination stress (NBIS).

ITZO and IGZO are ~0.05 and ~0.15 eV, respectively. The donor centration in ITZO is required to be suppressed by less than 2 orders of magnitude than IGZO to have $\sim 10^{16}$ cm^{-3} for appropriate TFT channel as shown in Figure 2.30c. These results provide experimental evidence that TAOS TFTs have the potential to replace LTPS-TFTs.

In a-IGZO, high mobility TFTs were reported using new layer structures or deposition processes. In general, the electrical properties of amorphous oxide semiconductors are strongly dependent on their chemical composition; the indium (In)-rich IGZO achieves a high mobility of 50 cm^2/V s. However, these TFTs possess an issue of negative threshold voltage owing to intrinsically high carrier density. Park et al. [49] reported In-rich IGZO

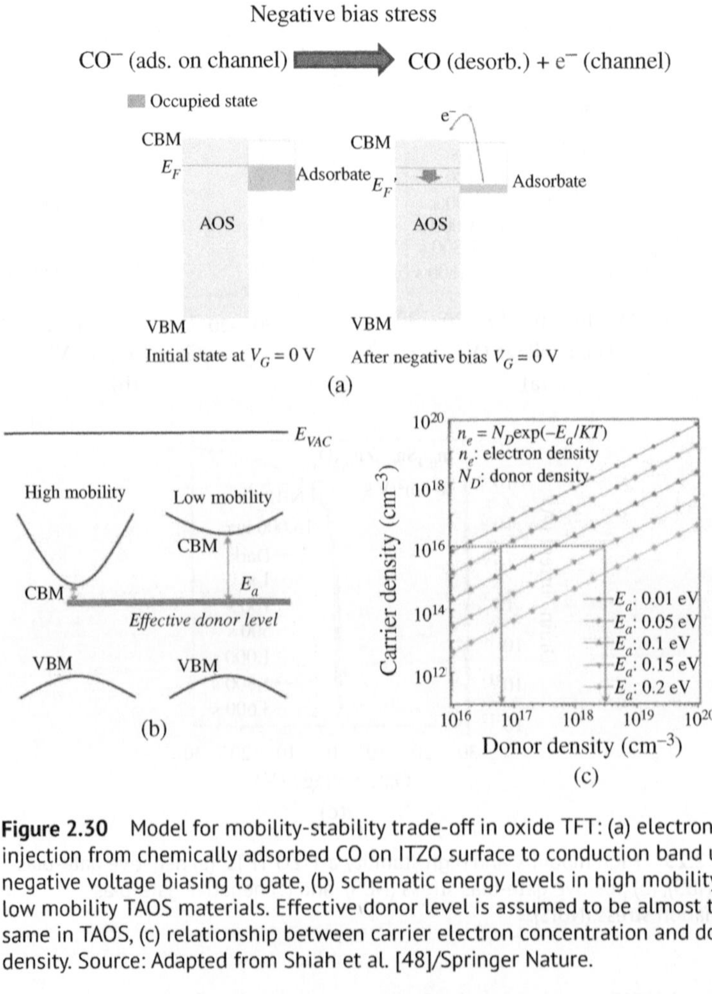

Figure 2.30 Model for mobility-stability trade-off in oxide TFT: (a) electron injection from chemically adsorbed CO on ITZO surface to conduction band upon negative voltage biasing to gate, (b) schematic energy levels in high mobility and low mobility TAOS materials. Effective donor level is assumed to be almost the same in TAOS, (c) relationship between carrier electron concentration and donor density. Source: Adapted from Shiah et al. [48]/Springer Nature.

TFTs with vertically stacked InO_x, ZnO_x, and GaO_x atomic layers by plasma-enhanced atomic layer deposition (PEALD) exhibited excellent performances such as saturation mobilities of ~74 cm^2/V s, threshold voltage of −1.3 V, I_{on}/I_{off} ratio of 10^8, *SS* of 0.26 V/decade, and hysteresis of 0.2 V, while keeping a reasonable carrier density of ~10^{17} cm^{-3}. High mobility TFTs were reported by the formation of a hetero-structured channel. For instance, a mobility of ~65 cm^2/V s was reported for bilayer TFTs of InGaZnO/In_2O_3 [50]. The authors explained this mobility enhancement in terms of compensation of oxygen vacancy in In_2O_3 layer with oxygen ions in the IGZO layer.

Utilization of quantum confinement effect is a possible strategy for mobility enhancement. Abe et al. [51] examined the quantum channel effects on TFT mobility using a combination of IGZO (channel, well) with a-Ga_2O_3(wall) and various thicknesses of IGZO. A clear confinement of electrons was observed in the blue shift of the band gap and plateaus in field-effect mobility vs. gate voltage relations when the well thickness was 5 nm or less. Device simulation incorporating the quantum confinement effects reproduced the well-thickness dependence of the plateau onset voltage, verifying that the quantized electronic levels in the a-IGZO well dominate the TFT characteristics and, thus, the electron transport is free electron-like with a coherent length of ~5 nm. However, the field-effect mobility did not exceed the drift mobility of bulk a-IGZO.

In_2O_3 has a higher Hall mobility but the electron carriers are easily created at high concentrations which are beyond the level for TFT channel application as in known as the parent oxides for ITO (In_2O_3:Sn), representative TCO. Thus, suppression of carrier electron generation is a key to applying In_2O_3 to TFT channels. High mobility In_2O_3-TFTs have been reported in several papers to date. For instance, mobility of >120 $cm^2/V\,s$ was reported for In_2O_3 thin films grown by ion-assisted deposition using an inorganic-organic hybrid gate insulator. In 2022, a very high mobility TFT (~140 $cm^2/V\,s$) was reported for In_2O_3:H using a conventional SiO_2/Si bottom-gate, as shown in Figure 2.31 [52]. The thin films were deposited by conventional sputtering in the atmosphere of $Ar + O_2 + H_2$. The carrier concentration was reduced to 10^{17} cm^{-3} by optimizing post-annealing temperature in this work. This mobility is higher by an order of magnitude than that in conventional In_2O_3 thin films deposited in a nominally H_2-free atmosphere and exceeds the Hall mobilities of the sample films covering 10^{17}~10^{20} cm^{-3}. Although instability is still observed for negative gate biasing, this report would be worth reexamining in detail, in particular elucidating the role of hydrogen appears to be critical for understanding this mobility enhancement.

2.4.2 Oxide Composition Effect on P-channel TFT Performance

Conduction in wide-gap oxide semiconductors is n-type in general, and p-type oxides are rather limited. A wide gap p-type oxide material, $CuAlO_2$, was reported in 1997 along with a design concept [4], but this material did not work well as p-channel TFTs like Cu_2O. The first p-channel oxide TFT was realized by the use of SnO [5], and although performance is not satisfactory, the first oxide C-MOS was reported by the use of SnO with ambipolarity [6].

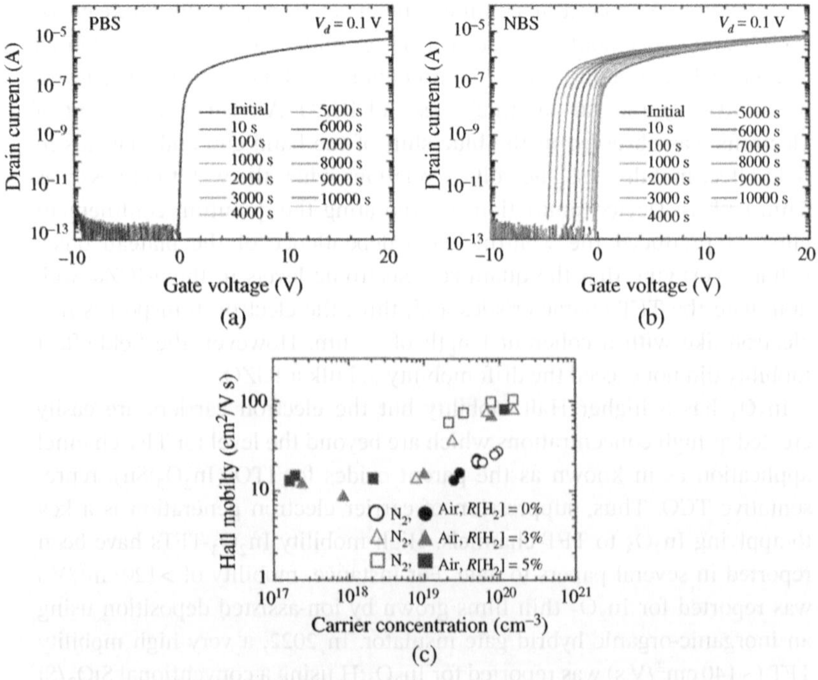

Figure 2.31 TFT performance and Hall mobility of In_2O_3:H (a) positive bias stability, (b) negative bias stability, (c) Hall mobility and carrier concentration in the thin films. Source: Magari et al. [52]/Springer Nature/CC-BY-4.0.

Oxides are unfavorable p-type semiconductors because of the deep energy level and the highly localized nature of O2p orbitals dominating the VBM. It is hard to fabricate p-channel TFTs with high mobilities. A practical approach to p-channel oxide TFTs is to use Cu^+ with $3d^{10}4s^0$ configuration and Sn^{2+} with $5s^2$ configuration. Cu_2O and SnO p-type materials and other p-type conductors bearing these cations in a variety of complex oxides have been reported. However, progress in p-channel TFT with channel of these oxides [53] is very slow compared with n-channel TFTs as show in Figure 2.32. SnO was the first reported ambipolar oxide and the first ambipolar TFT [6] was reported using this material. Since the Hall mobility (2–3 cm^2/V s) in epitaxial thin films is close to TFT mobility [55], this mobility appears to be upper limit for SnO. However, the situation of Cu_2O is totally different from SnO. Hall mobility in epitaxial thin films is very high (~90 cm^2/V s) but the TFT mobility using such a film remains 1–3 cm^2/(V s) [56]. Such a low mobility has not changed since the Shockley's report in 1938. Poor TFT mobility is not restricted but

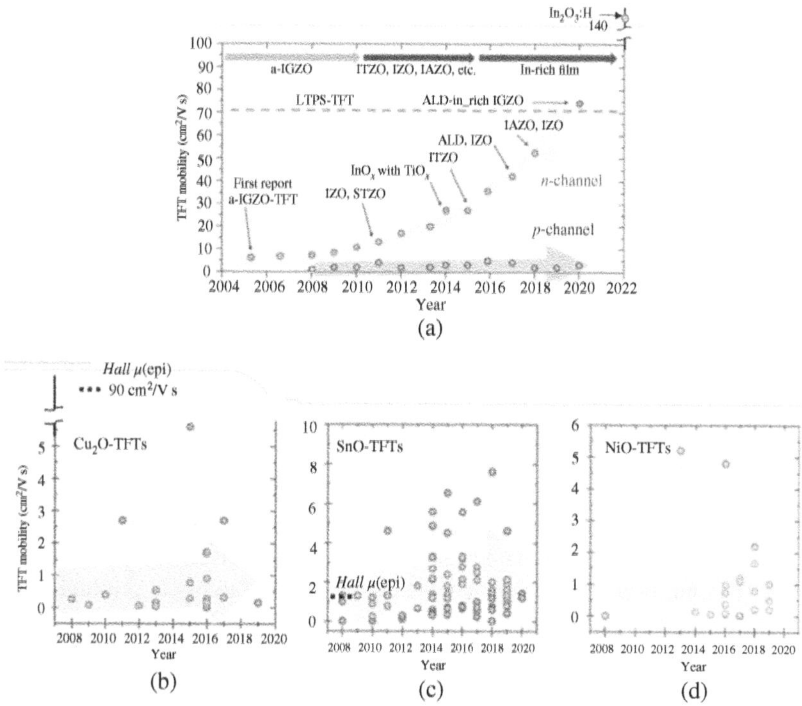

Figure 2.32 Improvement of mobility of (a) various types of oxide TFTs, (b) p-channel Cu_2O TFTs. Source: Adapted from Al-Jawhari [56], (c) p-channel Cu_2O TFTs, (d) p-channel SnO TFTs.

reported commonly for Cu^+-bearing complex oxide semiconductors. The main cause is likely to easy oxidation of Cu ion from +1 to +2 on the top surface region. If this idea is correct, p-type semiconductors composed of less electro-negative anions combined with Cu^+ should work well as p-channel TFTs. In practice, CuI-TFTs with high mobility were reported to date [57, 58]. However, reproducibility and stability of these TFTs would need to be verified for application.

P-channel oxide TFTs using β-TeO_2 was recently reported [9]. The sample was bi-layered nanosheet (1.5–2 nm thick) fabricated through surface oxidation of a eutectic mixture of tellurium and selenium. Figure 2.33 shows the local structure of β-TeO_2 and TFT data. Tellurium with +4 charge state takes distorted tetrahedral coordination and has $5s^2$ lone pair electrons. Such a local structure and $5s^2$ lone pair electron of Te^{4+} is almost the same as Sn^{2+} in SnO. It is thus considered the valence band top is primarily composed of percolated Te 5s orbitals as illustrated in the figure. The ionization potential

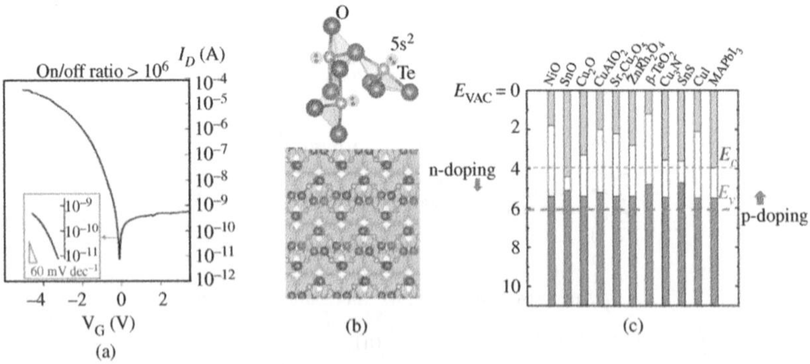

Figure 2.33 Bilayer β-TeO_2 TFT. (a) Transfer curve [9], (b) top: local structure, bottom: crystal structure overlayed with Te 5s orbital spread), (c) band alignment of p-type oxide semiconductors and empirical doping window.

of β-TeO_2 is ~5 eV, which is distinctly smaller than the empirical hole-doping limit (~6 eV) and the energy of CBM is the highest among p-type oxide semiconductors reported to date as summarized in Figure 2.33. The estimated TFT mobility at RT is very high, ~230 $cm^2/V\,s$, and the I_{on}/I_{off} ratio is 10^6. However, the effective hole mass evaluated by DFT calculation is ~0.5, not so small among oxide semiconductors. Since it was pointed out that evaluation of TFT mobility for nanosheet is difficult and has several pitfalls [59], further effort appears to be required to examine the validity of TFT performance. Te is easily reduced to Te^0 by thermal or electrochemical treatment. Since Te is a p-type semiconductor, caution is required.

2.4.3 Role of Hydrogen Impurity on Oxide TFT Characteristics

Hydrogen is one of the most common impurities in semiconductors. The critical roles of hydrogen on electrical properties have dual faces, favorable and unfavorable roles. An example of the former is hydrogen in a-Si:H, where silicon dangling bonds are passivated effectively with H forming Si—H bonds, making it possible to control p/n by impurity doping [53]. The latter example is seen in GaN where impurity hydrogen incorporated during thin film deposition by chemical vapor deposition works as charge compensators of hole [60]. Oxides contain hydrogen species as an impurity irrespective of crystal and amorphous state. This fact comes from strong affinity of oxygen to hydrogen.

2.4.3.1 State and Concentration of Hydrogen

Amorphous oxide semiconductor (AOS) thin films often contain high-density impurity hydrogen, e.g., 10^{20-21} cm^{-3}, which was considered

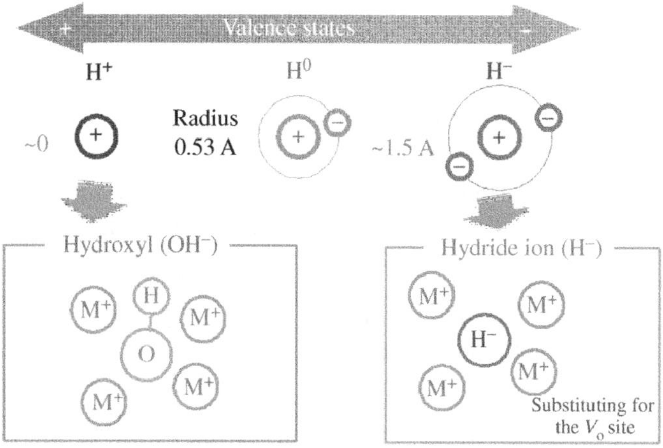

Figure 2.34 Three types of hydrogen in solids. No stable H^0 at room temperature has been reported in AOS materials.

to be mainly in the form of hydroxyl (OH) groups in early 2010s. Since thermal annealing around 300–400 °C is required to stabilize AOS-TFTs, it is natural to suggest a close relation between the stabilization/desorption of these hydrogen impurities and the TFT electrical properties such as mobility and subthreshold swing. In 2017, Bang et al. [61] found hydrogen directly bonding to metal cations as a hidden form of impurity hydrogens and their concentration is comparable to that of OH groups. Figure 2.34 summarizes three representative charge states and the chemical structures of each hydrogen in solids. In ionic solids such as oxides, neutral state of hydrogen is unstable and is stabilized in terms of charged states, H^+ and H^-. The chemical state of hydrogen is thus restricted to H^+ (OH group) and H^- (M–H) in AOS materials.

Figure 2.35 shows the depth concentration of H measured by secondary-ion mass spectroscopy (SIMS) in a-IGZO thin films sputter-deposited in the conventional vacuum chamber (base pressure $\sim 10^{-4}$ Pa, denoted standard "STD") [62]. The H-concentrations in the as-deposited thin films is up to $\sim 5 \times 10^{20}\ cm^{-3}$ throughout the sample. The reduction of the concentration is restricted to $\sim 5 \times 10^{19}\ cm^{-3}$ even in deposited in ultrahigh vacuum (UHV, $\sim 10^{-7}$ Pa) chamber. Thermal desorption spectra (TDS) of the as-deposited thin films reveal the desorbed gases are H_2, O_2, and H_2O, and each of them starts to desorb from ~100 °C, the desorption distinctly increases up to >350 °C and reaches a maximum at ~450 °C for H_2, ~500 °C for O_2, and ~450 °C for H_2O. It is noted that the desorption of Zn is distinct compared with In and Ga, and its TDS profile is similar to that of H_2O, suggesting

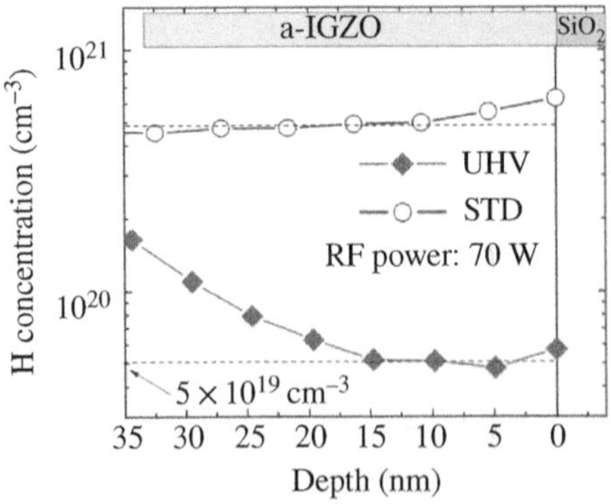

Figure 2.35 SIMS depth concentration of H in a-IGZO-thin films deposited on SiO_2/Si substrate. Sputtering was performed at the same conditions except for back pressures (UHV:~10^{-7} Pa and STD:~10^{-4} Pa). STD corresponds to conventionally adopted conditions. Source: Miyase et al. [62]/IOP Publishing.

that thermal dehydration accompanies evaporation of Zn. Since the main residual gas species in high vacuum chambers evacuated with turbo molecular pumps are H_2 and H_2O, these results may be understood in terms of these hydrogen species are incorporated during the sputter deposition. Rather high temperature state is realized during sputtering, thereby high density of oxygen vacancy is created as a result of the formation of low metal valence state at high temperatures. It is likely to occur that these oxygen vacancies react readily with residual H_2 and H_2O to give rise to OH and M–H.

Figure 2.36a shows infrared absorption spectra of a-IGZO-thin films deposited on NaCl single crystal substrates which are transparent to infrared region focused here [61]. There are two H-associated bands, a broad absorption around ~3000 cm^{-1} and three sharp bands located in the range ~1000–1500 cm^{-1}. The former is attributed to strongly hydrogen-bonded OH. The latter three bands are assigned to In—H, Ga—H, and Zn—H bonds, referring to the previously reported data on respective metal hydrides and DFT calculations. Figure 2.36b summarizes the variation in the total hydrogens and each hydrogen species desorbed upon heating the as-deposited thin films [61]. Since the total hydrogen impurities in a-IGZO thin films sputter-deposited in the standard condition is ~5×10^{20} cm^{-3} and the M–H concentration and O–H were estimated to be ~2×10^{20} and ~3×10^{20} cm^{-3}, respectively.

If this analogy of amorphous SiO_2 is valid to AOS, the precursor of OH would be strained M–O–M, and OH is created through the reaction with H_2O and M–H is created by reactions of an oxygen vacancy coordinated by M cations with H_2. The numbers of the precursor strained M—O—M bond and the oxygen vacancy are evaluated to be ~3×10^{20} cm^{-3} and ~1×10^{20} cm^{-3}, respectively.

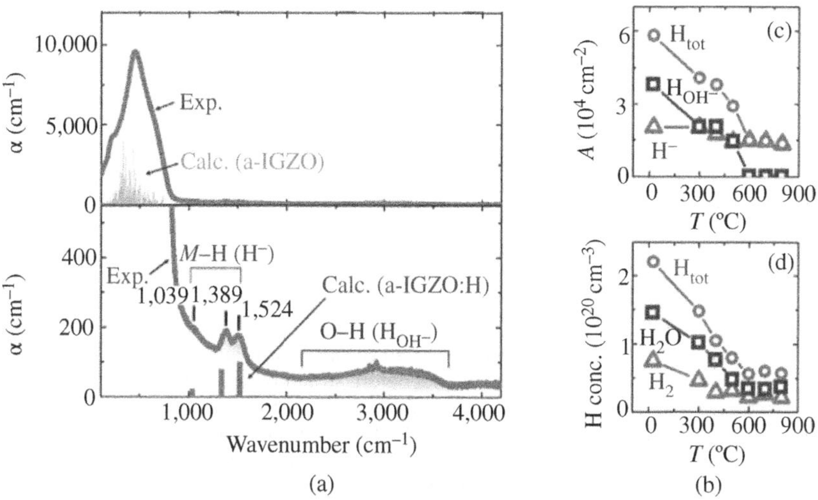

Figure 2.36 (a) IR spectra of a-IGZO thin film and variation in content H species with heating temperature, (b) (Top) O–H and M–H contents in terms of IR absorption area and (bottom) thermally desorbed H_2 and H_2O content, α: absorption coefficient, A: absorption area. Source: Bang et al. [61]/AIP Publishing.

The formation model for M–H by reaction of H_2 with neutral oxygen vacancy (NOV, trapping 2 electrons at the oxygen vacancy) was justified by a computation. Bang et al. [61] examined what happens if NOV and H_2 coexist in an a-IGZO structure mode [63] obtained by MD simulation. Local structure around the NOV is generated by classical MD (molecular dynamics) simulation on an oxygen-deficient a-IGZO model and an H_2 molecule is placed nearby the NOV. The H_2 reacts easily with the NOV to split to a pair of hydrogens doubly and triply coordinated with metal cations as shown in Figure 2.37, demonstrating the validity of the oxygen vacancy precursor model for the M—H bond. The validity of the strained M–O–M is discussed in the next section. Please note that this reaction eliminates the NOV without generating electron carrier. Li et al. [64] reported the M–H–M bridges may cause negative-bias illumination-stress instability (NBIS) in AOS-TFTs.

2.4.3.2 Carrier Generation and Hydrogen

Roles of hydrogen in AOS materials are not simple. In this section, the relation between hydrogen and carrier is described by separating low temperature and high temperature.

First is carrier generation by H-injection at low temperatures. Passivation of dangling bonds in a-Si:H makes it possible to control Fermi-level by

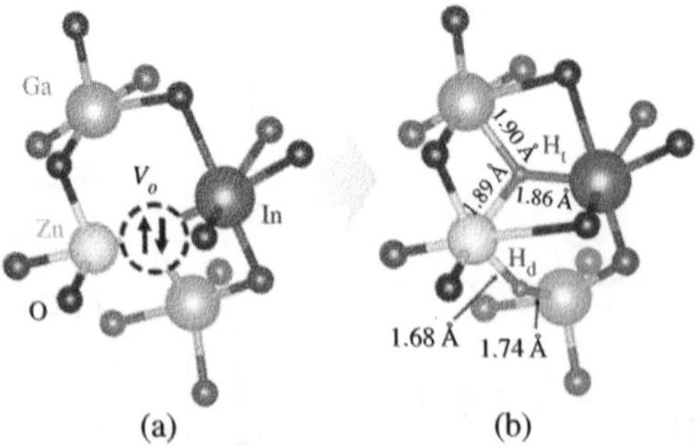

Figure 2.37 Reaction of neutral oxygen vacancy with H_2 molecule in a-IGZO: (a) a MD-simulated oxygen vacancy model before structural relaxation used as an initial structure for hydrogen containing a-IGZO, (b) local structure of hydrogen incorporated a-IGZO. H_t and H_d indicate the triply and doubly coordinated hydrogen atoms, respectively. This reaction of neutral oxygen vacancy with H_2 occurs readily if H_2 is present near V_o. Source: Bang et al. [61]/AIP Publishing.

impurity doping like crystalline Si although efficiency is much lower. At the early stage of AOS-TFTs, H-containing plasma processing was applied to deposition of gate insulators, but the AOS became so conducting that TFT cannot switch off due to the high carrier generation. This result is easily predicted by remembering the early research on AOS by proton implantation at room temperature (RT) [65]. On the other hand, no carrier generation occurs by exposing a-SiO_2 and a-Ga_2O_3 to H-containing plasma. This conspicuous difference in carrier generation by hydrogen may be understood in terms of the concept of universal charge transition level of hydrogen in compound semiconductors proposed by Walle [66]; the transition energy, $E(H^+/H^-)$ from H^- to H^+ is widely present at ~4.5 eV below the vacuum level irrespective of materials. This level is pretty close to the potential (−4.45 eV from vacuum level) of proton reduction to molecular hydrogen. Figure 2.38 summarizes the band edge energy of various AOS materials and a-SiO_2 along with the universal level of $E(H^+/H^-)$. When CBM is lower than $E(H^+/H^-)$, H is ionized to $H^+ + e^-$ and the electron and H^+ are stabilized by transferring e^- to the CBM and forming OH through attaching to oxygen ions. The experimental results agree with this prediction. Conventional wide-gap amorphous oxide insulators represented by a-SiO_2 have much smaller electron affinity (higher CBM). Thus, no carrier generation does not occur in these materials even by exposing H-plasma or implantation.

Second is relation between carrier concentration and hydrogen in a-IGZO upon thermal annealing after deposition. This relation is rather complex

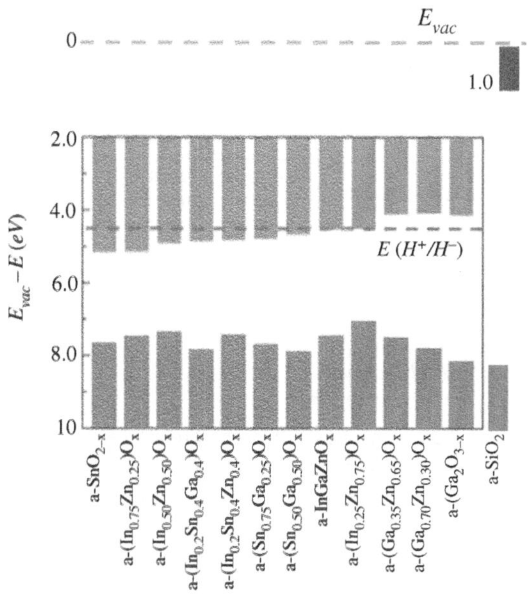

Figure 2.38 Energy levels of VBM and CBM of various AOS materials along with and hydrogen charge transition level (4.5 eV). Data of a-SiO_2 are shown for comparison. Source: Hosono and Kamiya [67]/John Wiley & Sons.

described below and further effort is needed to get a comprehensive understanding at the present stage.

Figure 2.39 summarizes the features of the various a-IGZO thin films which were deposited at different conditions and annealed under different atmospheres. Hydrogen impurities are incorporated at surprisingly high concentrations ($>10^{20}$ cm^{-3}) for AOS thin films sputter-deposited at standard conditions (STD, base pressure $\sim 10^{-4}$ Pa). This concentration can be reduced by an order of magnitude by deposition in higher vacuum conditions (UHV, $\sim 10^{-7}$ Pa). The chemical forms of impurity hydrogen are OH and M–H, and each of them has comparable concentrations. Carrier concentrations in the former STD thin films are $\sim 10^{16}$ cm^{-3} which works as the TFT channel, but the latter UHV thin films are not suitable due to too low carrier concentrations. Orui et al. [68] proposed this carrier reduction was ascribed to the presence of excess oxygen which compensates the shallow donor from an observation of distinct reduction of the deep V_o giving the subgap absorption above the VBM. The UHV chamber contains less residual H_2 concentration, thereby the conventionally used partial pressure of O_2 gas (P_{O2}) appropriate to fabricate AOS thin films for TFTs produces excess O_2 pressure because of less consumption of O_2 due to less formation of H_2O by reaction with residual H_2. These results may be understood by considering the sputtering process. The gas plasma state and collision of sputtering ions with an oxide ceramic target create high

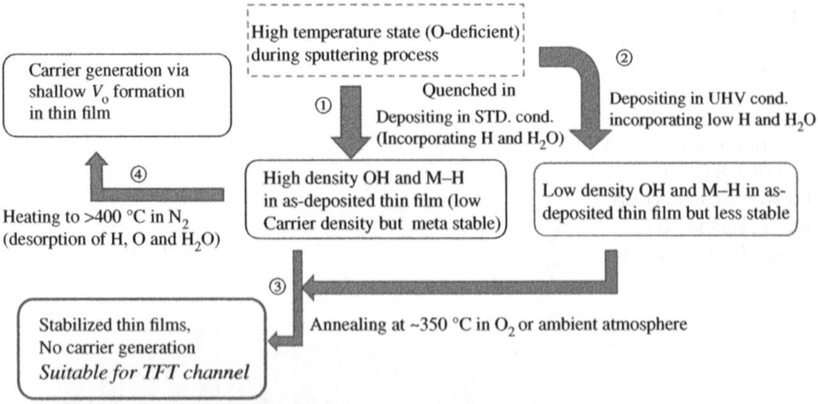

Figure 2.39 Changes in hydrogen-related species and carrier density in AOS thin films with deposition and post annealing. Source: Hosono and Kamiya [67]/John Wiley & Sons.

temperature state in which oxygen deficient states are stabilized. Since sputter-deposition is an extremely rapid quenching process (quenching from the high-temperature gas/high-energy precursor states to the stabilized states at the growing surface), the resulting thin films have highly strained M–O–M and oxygen vacancies. Hydrogen and water are the main residual gas species in the standard vacuum chamber. Water molecule reacts with the strained M—O—M bonds to form OH, while H_2 does with the oxygen vacancy to give M—H bonds.

When the as-deposited STD thin films are heated to >400 °C in an N_2 atmosphere, the carrier concentration increases to $\sim 10^{20}$ cm^{-3}, which is too high to be applied to a TFT channel. This carrier generation originates from the shallow oxygen vacancy (V_o) which is created by the desorption of H_2O and H_2 from OH and M–H following the reactions (2.1) and (2.2)

$$\text{M–OH} + \text{H–M} = \text{M–}V_o\text{–M} + \text{H}_2\text{O} \tag{2.1}$$

$$\text{M–H} + \text{H–M} = \text{M–}V_o\text{–M} + \text{H}_2 \tag{2.2}$$

When the as-deposited thin films are annealed in an O_2 atmosphere around ~350 °C, carrier generation is suppressed and structurally stabilized thin films are obtained. Since wet O_2 atmosphere promotes the oxygen diffusion/oxidation and structural relaxation during annealing than dry O_2 annealing, annealing at an ambient atmosphere (note it includes H_2O) works well to produce good a-IGZO thin films for TFT channels [29]. However, we should note that the wet O_2 annealing incorporates extra H into the a-IGZO films; therefore, it is not conformably concluded that wet O_2 annealing is better than dry O_2 annealing in terms of TFT stability.

As described above, the role of hydrogen in oxide TFTs is complex and can either favor or harm electrical properties. The effect is not determined by total concentration depending on the chemical state, OH and M–H. In addition, the post annealing induces the concentration changes of not only total hydrogen but also the ratio of these two H species. As a consequence, carrier electron concentration is changed via V_0 formation by (2.1) and (2.2) and V_0 annihilation through the reaction of V_0 with H, as shown in Figure 2.39. When the deposition process is changed, it is natural to consider the concentration of impurity hydrogen strongly depends on the deposition process.

2.5 Summary

This chapter described the material characteristics of oxide semiconductors in comparison with conventional semiconductors based on the diamond-type structure. The points of this chapter are summarized below:

1. Chemical bonding in oxide semiconductors is dominated by ionic nature, which totally differs from the conventional semiconductors based on covalent bonding.
2. The valence band maximum of oxide semiconductors is primarily composed of oxygen 2p-orbitals which are localized, while the cation's vacant orbitals which are spatially spread constitute the conduction band minimum. Thus, oxide semiconductors are mainly of n-type and large electron mobility materials are obtained in a post-transition system with $(n-1)d^{10}ns^0$ $(n \geq 4)$ configuration. P-type materials are realized by strategy for designing the delocalized state at the valence band maximum but materials that work well as p-channel semiconductors are rather limited until now.
3. P/N-orientation and carrier dopability may be understood based on bandalignment. N-type doping is possible when electron affinity is below ~4 eV, and materials with ionization potential smaller than ~6 eV are plausible candidates for P-type. Please note this is an empirical rule and easiness of counter doping such as electron for p-type competes with this rule.
4. Amorphous oxide semiconductors with large electron mobility comparable to polycrystals are obtained in systems of a post-transition metal cation with $4d^{10}5s^0$ configuration. This situation differs from the conventional semiconductors with diamond structure and originates from structure-insensitive metal s-orbital percolation at the conduction band minimum.

5. Oxide semiconductors contain not a negligibly small amount of hydrogen species, in particular thin films deposited. They exist in two forms, i.e., O–H and M–H. These species are associated with carrier generation/annihilation.
6. Issues for display and memory application were described along with the current status.

References

1 Robertson, J. (1979). Electronic structure of SnO_2, GeO_2, PbO_2, TeO_2 and MgF_2. *Journal of Physics C: Solid State Physics* 12 (22): 4767.

2 Orita, M., Ohta, H., Hirano, M. et al. (2001). Amorphous transparent conductive oxide $InGaO_3$ (ZnO) m (m $\leq$ 4): a Zn4s conductor. *Philosophical Magazine B* 81 (5): 501–515.

3 Kawazoe, H., Yanagi, H., Ueda, K. et al. (2000). Transparent p-type conducting oxides: design and fabrication of pn heterojunctions. *MRS Bulletin* 25 (8): 28–36.

4 Kawazoe, H., Yasukawa, M., Hyodo, H. et al. (1997). P-type electrical conduction in transparent thin films of $CuAlO_2$. *Nature* 389 (6654): 939–942.

5 Ogo, Y., Hiramatsu, H., Nomura, K. et al. (2009). Tin monoxide as an s-orbital-based p-type oxide semiconductor: electronic structures and TFT application. *Physica Status Solidi (A)* 206 (9): 2187–2191.

6 Nomura, K., Kamiya, T., and Hosono, H. (2011). Ambipolar oxide thin-film transistor. *Advanced Materials* 23 (30): 3431–3434.

7 Wang, Z., Nayak, P.K., Caraveo-Frescas, J.A. et al. (2016). Recent developments in p-type oxide semiconductor materials and devices. *Advanced Materials* 28 (20): 3831–3892; Lee, A.W., Zhang, Y., Huang, C.H. et al. (2020). Switching mechanism behind the device operation mode in SnO-TFT. *Advanced Electronic Materials* 6 (12): 2000742.

8 Walsh, A. and Watson, G.W. (2005). The origin of the stereochemically active Pb (II) lone pair: DFT calculations on PbO and PbS. *Journal of Solid State Chemistry* 178 (5): 1422–1428.

9 Zavabeti, A., Aukarasereenont, P., Tuohey, H. et al. (2021). High-mobility p-type semiconducting two-dimensional β-TeO_2. *Nature Electronics* 4 (4): 277–283.

10 Mizoguchi, H., Hirano, M., Fujitsu, S. et al. (2002). $ZnRh_2O_4$: a p-type semiconducting oxide with a valence band composed of a low spin state of Rh_3^+ in a $4d^6$ configuration. *Applied Physics Letters* 80 (7): 1207–1209.

11 Kamiya, T., Narushima, S., Mizoguchi, H. et al. (2005). Electrical properties and structure of p-type amorphous oxide semiconductor xZnO·Rh_2O_3. *Advanced Functional Materials* 15 (6): 968–974.

12 Narushima, S., Mizoguchi, H., Shimizu, K.I. et al. (2003). A p-type amorphous oxide semiconductor and room temperature fabrication of amorphous oxide p–n heterojunction diodes. *Advanced Materials* 15 (17): 1409–1413.

13 Schein, F.L., Winter, M., Böntgen, T. et al. (2014). Highly rectifying p-$ZnCo_2O_4$/n-ZnO heterojunction diodes. *Applied Physics Letters* 104 (2): 022104.

14 Hosono, H. (2013). Exploring electro-active functionality of transparent oxide materials. *Japanese Journal of Applied Physics* 52 (9R): 090001.

15 Hosono, H. and Kitano, M. (2021). Advances in materials and applications of inorganic electrodes. *Chemical Reviews* 121 (5): 3121–3185.

16 Hosono, H., Ogo, Y., Yanagi, H. et al. (2010). Bipolar conduction in SnO thin films. *Electrochemical and Solid-State Letters* 14 (1): H13.

17 Burbano, M., Scanlon, D.O., and Watson, G.W. (2011). Sources of conductivity and doping limits in CdO from hybrid density functional theory. *Journal of the American Chemical Society* 133 (38): 15065–15072.

18 Zunger, A. (2003). Practical doping principles. *Applied Physics Letters* 83 (1): 57–59.

19 Street, R.A. (1991). *Hydrogenated Amorphous Silicon*. Cambridge University Press.

20 Spear, W.E. and Le Comber, P.G. (1975). Substitutional doping of amorphous silicon. *Solid State Communications* 17 (9): 1193–1196.

21 Le Comber, P.G., Spear, W.E., and Ghaith, A. (1979). Amorphous-silicon field-effect device and possible application. *Electronics Letters* 6 (15): 179–181.

22 Hosono, H., Yasukawa, M., and Kawazoe, H. (1996). Novel oxide amorphous semiconductors: transparent conducting amorphous oxides. *Journal of Non-Crystalline Solids* 203: 334–344.

23 Wager, J.F., Yeh, B., Hoffman, R.L. et al. (2014). An amorphous oxide semiconductor thin-film transistor route to oxide electronics. *Current Opinion in Solid State and Materials Science* 18 (2): 53–61.

24 Hosono, H., Kikuchi, N., Ueda, N. et al. (1996). Working hypothesis to explore novel wide band gap electrically conducting amorphous oxides and examples. *Journal of Non-Crystalline Solids* 198: 165–169.

25 Nomura, K., Ohta, H., Takagi, A. et al. (2004). Room-temperature fabrication of transparent flexible thin-film transistors using amorphous oxide semiconductors. *Nature* 432 (7016): 488–492.

26 Hosono, H. (2006). Ionic amorphous oxide semiconductors: material design, carrier transport, and device application. *Journal of Non-crystalline Solids* 352 (9–20): 851–858.

27 Narushima, S., Hosono, H., Jisun, J. et al. (2000). Electronic transport and optical properties of proton-implanted amorphous $2CdO{\cdot}GeO_2$ films. *Journal of Non-Crystalline Solids* 274 (1–3): 313–318l.

28 Narushima, S., Orita, M., Hirano, M. et al. (2002). Electronic structure and transport properties in the transparent amorphous oxide semiconductor $2CdO{\cdot} GeO_2$. *Physical Review B* 66 (3): 035203.

29 Kamiya, T., Nomura, K., and Hosono, H. (2010). Present status of amorphous In–Ga–Zn–O thin-film transistors. *Science and Technology of Advanced Materials* 11: 044305–044328.

30 Kamiya, T., Nomura, K., and Hosono, H. (2009). Electronic structures above mobility edges in crystalline and amorphous In-Ga-Zn-O: percolation conduction examined by analytical model. *Journal of display technology* 5 (12): 462–467.

31 Nenashev, A.V., Oelerich, J.O., Greiner, S.H.M. et al. (2019). Percolation description of charge transport in amorphous oxide semiconductors. *Physical Review B* 100 (12): 125202.

32 Hossain Chowdhury, M.D., Migliorato, P., and Jang, J. (2013). Low temperature characteristics in amorphous indium-gallium-zinc-oxide thin-film transistors down to 10 K. *Applied Physics Letters* 103 (15): 152103.

33 Nomura, K., Kamiya, T., Hirano, M. et al. (2009). Origins of threshold voltage shifts in room-temperature deposited and annealed a-In–Ga–Zn–O thin-film transistors. *Applied Physics Letters* 95 (1): 013502.

34 Kimura, M., Nakanishi, T., Nomura, K. et al. (2008). Trap densities in amorphous-$InGaZnO_4$ thin-film transistors. *Applied Physics Letters* 92 (13): 133512.

35 Street, R.A. (1982). Doping and the fermi energy in amorphous silicon. *Physical Review Letters* 49 (16): 1187.

36 Kamiya, T. and Hosono, H. (2010). Material characteristics and applications of transparent amorphous oxide semiconductors. *NPG Asia Materials* 2 (1): 15–22.

37 Ide, K., Nomura, K., Hosono, H. et al. (2019). Electronic defects in amorphous oxide semiconductors: a review. *Physica Status Solidi (a)* 216 (5): 180037.

38 Nomura, K., Kamiya, T., Ikenaga, E. et al. (2011). Depth analysis of subgap electronic states in amorphous oxide semiconductor, a-In-Ga-Zn-O, studied by hard x-ray photoelectron spectroscopy. *Journal of Applied Physics* 109 (7): 073726.

39 Kobayashi, K., Yabashi, M., Takata, Y. et al. (2003). High resolution-high energy x-ray photoelectron spectroscopy using third-generation synchrotron radiation source, and its application to Si-high k insulator systems. *Applied Physics Letters* 83 (5): 1005–1007.

40 Ryu, B., Noh, H.K., Choi, E. et al. (2010). O-vacancy as the origin of negative bias illumination stress instability in amorphous In–Ga–Zn–O thin film transistors. *Applied Physics Letters* 97 (2): 022108.

41 Oh, H., Park, S.H.K., Ryu, M.K. et al. (2012). Improved stability of atomic layer deposited ZnO thin film transistor by intercycle oxidation. *ETRI Journal* 34 (2): 280–283.

42 Hosono, H. (2022). Chapter 1. In: *Amorphous Oxide Semiconductors: IGZO and Related Materials for Display and Memory* (ed. H. Hosono and H. Kumomi). New York: Wiley.

43 Kamiya, T., Nomura, K., and Hosono, H. (2010). Origin of definite Hall voltage and positive slope in mobility-donor density relation in disordered oxide semiconductors. *Applied Physics Letters* 96 (12): 122103.

44 Ide, K., Nomura, K., Hiramatsu, H. et al. (2012). Structural relaxation in amorphous oxide semiconductor, a-In-Ga-Zn-O. *Journal of Applied Physics* 111 (7): 073513.

45 Yeon, H.W., Lim, S.M., Jung, J.K. et al. (2016). *NPG Asia Materials* 8 (3): e250–e250.

46 Matsueda, Y. (2010). Required characteristics of TFTs for next generation flat panel display backplanes, *ITC'10*: 314–317.

47 Shiah, Y.S., Sim, K., Ueda, S. et al. (2021). Unintended carbon-related impurity and negative bias instability in high-mobility oxide TFTs. *IEEE Electron Device Letters* 42 (9): 1319–1322.

48 Shiah, Y.S., Sim, K., Shi, Y. et al. (2021). Mobility–stability trade-off in oxide thin-film transistors. *Nature Electronics* 4 (11): 800–807.

49 Sheng, J., Hong, T., Lee, H.M. et al. (2019). Amorphous IGZO TFT with high mobility of~ 70 $cm^2/(Vs)$ via vertical dimension control using PEALD. *ACS Applied Materials & Interfaces* 11 (43): 40300–40309.

50 He, J., Li, G., Lv, Y. et al. (2019). Defect self-compensation for high-mobility bilayer InGaZnO/In_2O_3 thin-film transistor. *Advanced Electronic Materials* 5 (6): 1900125.

51 Abe, K., Kamiya, T., and Hosono, H. (2022). Quantum confinement effects in amorphous In–Ga–Zn–O thin-film transistors with quantum well channel. *Applied Physics Letters* 121 (22): 222104.

52 Magari, Y., Kataoka, T., Yeh, W. et al. (2022). High-mobility hydrogenated polycrystalline In_2O_3 (In_2O_3: H) thin-film transistors. *Nature Communications* 13 (1): 1078.

53 Wagner, S. (2010). Amorphous silicon: vehicle and test bed for large-area electronics. *Physica Status Solidi a* 207 (3): 501–509.

54 Nomura, K. (2021). Recent progress of oxide-TFT-based inverter technology. *Journal of Information Display* 22 (4): 211–229.

55 Ogo, Y., Hiramatsu, H., Nomura, K. et al. (2008). P-channel thin-film transistor using p-type oxide semiconductor. *SnO. Applied Physics Letters* 93 (3): 032113.

56 Al-Jawhari, H.A. (2015). A review of recent advances in transparent p-type Cu_2O-based thin film transistors. *Materials Science in Semiconductor Processing* 40: 241–252.

57 Liu, A., Zhu, H., Park, W.T. et al. (2020). High-performance p-channel transistors with transparent Zn doped-CuI. *Nature Communications* 11 (1): 4309.

58 Lee, S., Lee, H.J., Ji, Y. et al. (2020). Vacancy engineering of a solution processed CuI semiconductor: tuning the electrical properties of inorganic p-channel thin-film transistors. *Journal of Materials Chemistry C* 8 (28): 9608–9614.

59 Choi, H.H., Cho, K., Frisbie, C.D. et al. (2018). Critical assessment of charge mobility extraction in FETs. *Nature Materials* 17 (1): 2–7.

60 Nakamura, S., Mukai, T., Senoh, M. et al. (1992). Thermal annealing effects on p-type Mg-doped GaN films. *Japanese Journal of Applied Physics* 31 (2B): L19.

61 Bang, J., Matsuishi, S., and Hosono, H. (2017). Hydrogen anion and subgap states in amorphous In-Ga-Zn-O thin films for TFT applications. *Applied Physics Letters* 110 (23): 232105.

62 Miyase, T., Watanabe, K., Sakaguchi, I. et al. (2014). Roles of hydrogen in amorphous oxide semiconductor In-Ga-Zn-O: comparison of conventional and ultra-high-vacuum sputtering. *ECS Journal of Solid State Science and Technology* 3 (9): Q3085.

63 Kamiya, T., Nomura, K., Hirano, M. et al. (2008). Electronic structure of oxygen deficient amorphous oxide semiconductor a-$InGaZnO_{4-x}$: optical analyses and first-principle calculations. *Physica Status Solidi c* 5 (9): 3098–3100.

64 Li, H., Guo, Y., and Robertson, J. (2017). Hydrogen and the light-induced bias instability mechanism in amorphous oxide semiconductors. *Scientific Reports* 7 (1): 16858.

65 Hosono, H., Kikuchi, N., Ueda, N. et al. (1995). Amorphous transparent electroconductor $2CdO{\cdot}GeO_2$: conversion of amorphous insulating cadmium germanate by ion implantation. *Applied Physics Letters* 67 (18): 2663–2665.

66 Van de Walle, C.G. and Neugebauer, J. (2003). Universal alignment of hydrogen levels in semiconductors, insulators and solutions. *Nature* 423 (6940): 626–628.

67 Hosono, H. and Kamiya, T. (2022), Chapter 7). State and role of hydrogen in amorphous oxide semiconductors. In: *Amorphous Oxide Semiconductors* (ed. H. Hosono and H. Kumomi). New York: Wiley.

68 Orui, T., Herms, J., Hanyu, Y. et al. (2015). Charge compensation by excess oxygen in amorphous In-Ga-Zn-O films deposited by pulsed laser deposition. *Journal of Display Technology* 11 (6): 518–522.

3

Oxide Thin Film Transistor Devices

3.1 Introduction

The most important device of modern electronics is the field effect transistor (FET). This device is an important and growing part of the global economy enabling a large share of the entire global electronics market, as shown in Figure 3.1.

At the heart of a FET is a parallel plate capacitor formed by a metal plate separated from a semiconductor by a barrier layer, as shown in Figure 3.2.

For a p-type semiconductor substrate, a layer of conducting electrons might form at the metal/semiconductor interface. These electrons are attracted by the positive charge in the top plate. Likewise, for an n-type semiconductor substrate, if the voltage applied to the top metal contact is smaller than a threshold voltage, a layer of conducting holes forms at the metal/semiconductor interface. These holes are attracted by the negative charge in the top plate.

Julius E. Lilienfeld, [1, 2] an Austro-Hungarian-American physicist and electrical engineer, suggested providing contacts on the sides of the semiconductor plate, as shown in Figure 3.3a [2]. These contacts should be n^+ contacts for the n-channel device and p^+ contacts for the p-channel device, as shown in Figure 3.3a. A FET became a practical transistor only after pioneering work done by William Shockley in the early 1950s [4].

The capacitance between the gate and the channel at gate voltages above threshold is

$$C = \frac{\varepsilon_0 KWL}{d + \Delta d} \tag{3.1}$$

Here ε_0 is the dielectric permittivity of vacuum, K is the dielectric constant of the barrier layer, W and L are the channel width and length, respectively, d is the gate-to-channel separation, and Δd represents the capacitance associated with the conducting layer [3]. The conducting layer capacitance,

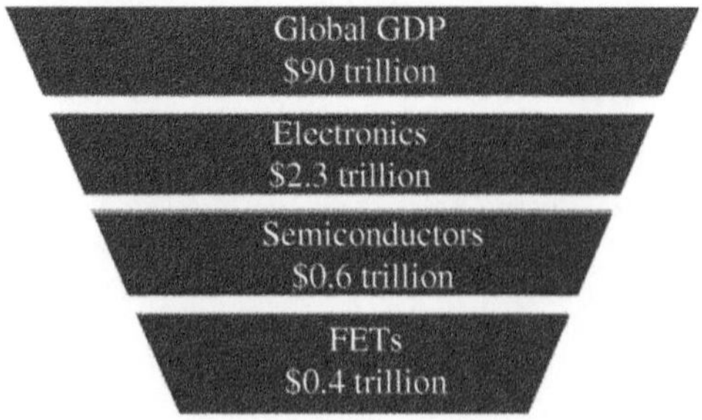

Figure 3.1 Estimates of global GDP and electronics, semiconductor, and FET markets for the year 2023.

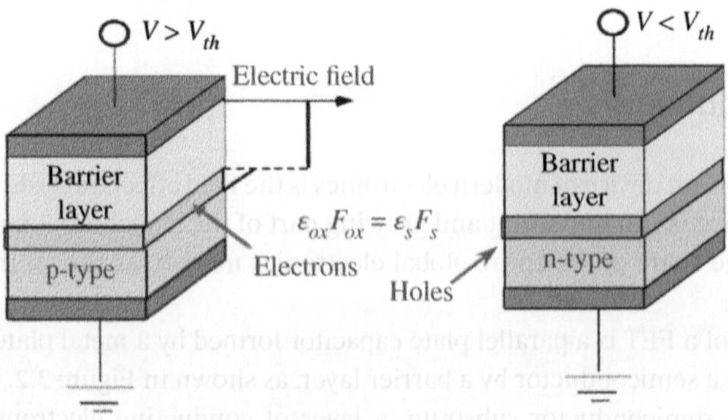

Figure 3.2 Metal-Insulator-Semiconductor (MIS) or Metal-Oxide-Semiconductor (MOS) capacitor. F_{ox}, F_s, and e_{ox}, e_s are electric fields and dielectric permittivities in the barrier layer and the semiconductor, respectively. V_{th} is threshold voltage; V is gate voltage.

which is connected in series with barrier layer capacitance is also referred to as "quantum capacitance" [5].

Above the threshold, the carrier concentration in the device channel is proportional to the gate voltage swing:

$$n_s = C(V - V_{th})/q$$
$$p_s = C(V_{th} - V)/q \qquad (3.2)$$

Here n_s and p_s are carrier densities in n-channel and p-channel FET, respectively. An output FET current flows between the drain and the source contacts and is controlled by the gate voltage swing ($\Delta V = V - V_{th}$ for n-channel FETs and $\Delta V = V_{th} - V$ for p-channel FETs). Below the threshold, a FET is in the off-state, with very few free carriers in the transistor channel between the source and the drain. Since the gate contact is isolated from the channel by the gate insulator, the input impedance is very high, a major advantage compared to bipolar junction transistors (BJTs).

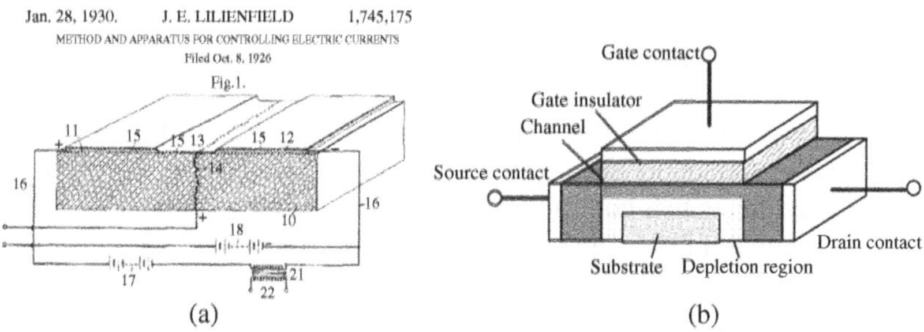

Figure 3.3 (a) Sketch of a FET from the original Lilienfeld patent [1], (b) schematic of a MOSFET. Source: Adapted from Shur [3].

FETs use different semiconductors and diverse ways of isolating the gate from the channel. The most important device by far is the silicon MOSFET, as shown in Figure 3.3b. In MOSFET, a native silicon oxide (SiO_2) has been used as a gate dielectric for older technology nodes. Nowadays, high-K dielectric layers (such as HfO_2 layers) are the key parts of the gate dielectric stack, but the acronym has not changed. MOSFETs can have an n-type channel (NMOS) or a p-type channel (PMOS). A MOSFET is actually a four-terminal device with source, gate, drain, and bulk (substrate) contact. The n^+ source and drain contacts and the n-type channel are isolated from the p-type substrate by the depletion regions formed in the same way as in a p-n junction. The substrate bias can control the width of these depletion regions.

Both n-channel and p-channel devices fabricated on the same substrate and sharing the same gate contact represent Complementary MOS (CMOS) technology. CMOS technology provides incredibly low power consumption because one of the n-channel and p-channel devices connected in series is off, except during switching the gate voltage. Figure 3.4a shows the circuit

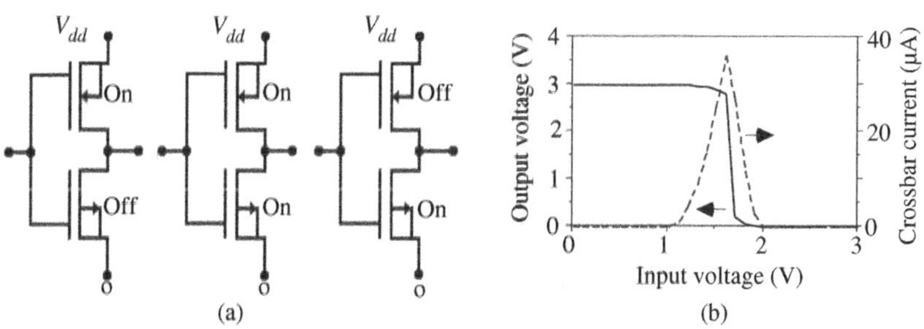

Figure 3.4 (a) Circuit symbols and (b) output voltage and current of a CMOS inverter. Source: Adapted from Shur [3].

of an inverter made of an n-channel transistor and a p-channel transistor at various switching states. Figure 3.4b shows the output characteristics of the inverter. When a positive voltage is applied to the gate of the circuit shown in Figure 3.4a, the n-type MOS is on, but the p-type MOS is off. When a negative office is applied to the gate, the p-type MOS is on, but the n-type MOS is off. The only time when both p-type and n-type transistors allow the current to flow between the positive V_{dd} terminal and ground is during switching. Hence, ideally, a CMOS inverter consumes energy only during switching. CMOS is the most important semiconductor technology. The integration scale for CMOS and MOSFETs has reached 2.6 trillion devices on a single chip [6] and continues to increase rapidly.

FETs implemented in other materials systems, such as GaAs, InGaAs, GaN, SiC, graphene, and many others compete with silicon for different niche applications, such as radio frequency (RF), power electronics, radiation hard electronics, and high-temperature electronics. The most

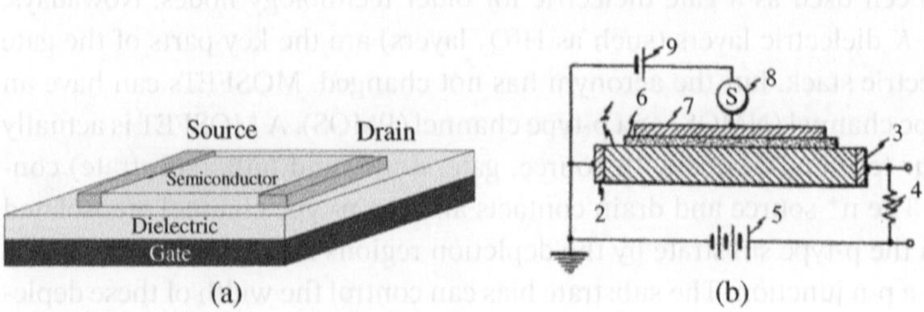

Figure 3.5 (a) Schematic TFT structure, (b) TFT figure from the 1959 TFT patent by J. T. Wallmark. Source: Adapted from Wallmark [7].

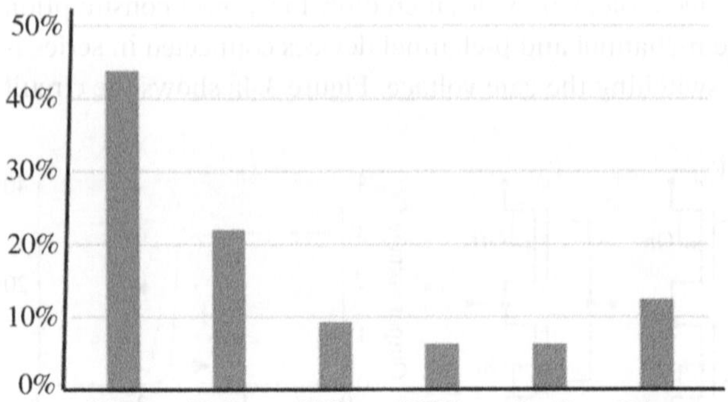

Figure 3.6 Relative use of oxide TFTs. Source: Adapted from Nomura [8].

important alternative to a MOSFET is a TFT, which is a FET made by thin semiconductor layer deposition fabricated on any substrate. Figure 3.5 shows (a) a schematic TFT structure and (b) a TFT from the 1959 pioneering TFT patent by John T. Wallmark of RCA [7].

Today a variety of oxide TFTs are explored, as shown in Figure 3.6 [8]. Oxide TFTs are poised to replace a-Si TFT as the dominant technology for flat panel displays. Detailed discussions on oxide TFT applications in displays and other products are included in Chapter 5 Applications. Compared to Si CMOS, oxide TFTs could operate at higher temperatures (up to 350 °C).

3.2 Oxide TFT Compared to a-Si, Poly-Si, Organic, and II-VI TFTs

Figure 3.7 shows the process temperature effect on the field effect mobility of oxide and other types of TFTs. Figure 3.8 compares the mobilities of various types of TFTs. The mobility gap between crystalline Si and TFTs is shrinking. As device dimensions shrink, intrinsic mobility plays a diminishing role because of the ballistic mobility effects due to contact scattering.

Oxide TFTs have reached deep submicron dimensions [9, 10] comparable with the mean free path of crystalline metal oxides. For example, for mobility of 345 $cm^2/V s$ for ZnO_x [11], we obtain 12 nm. This mobility value is approaching the values for short-channel Si CMOS [12]. As a

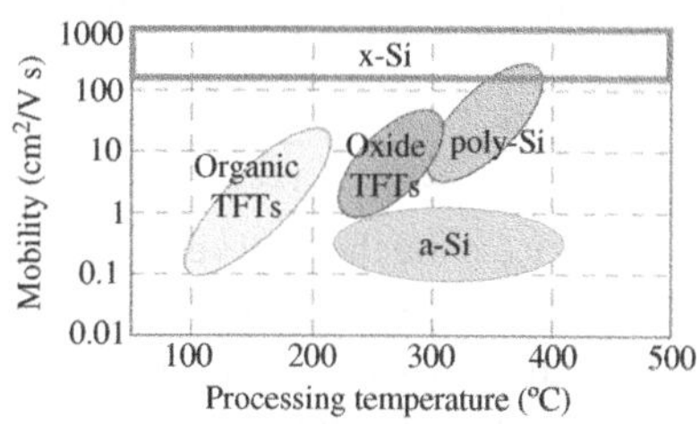

Figure 3.7 TFT field effect mobility vs. process temperature.

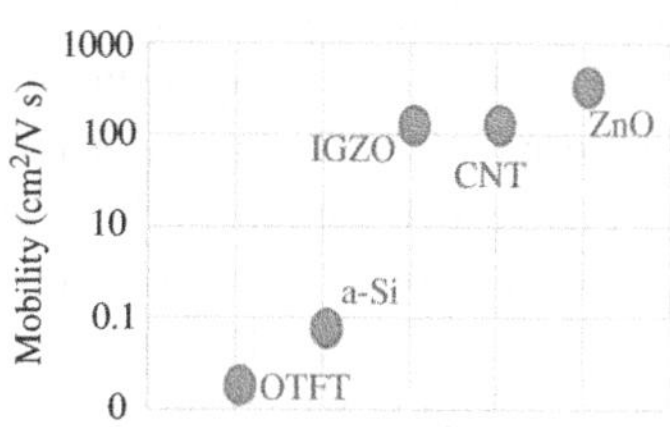

Figure 3.8 Comparison of mobilities of various TFTs.

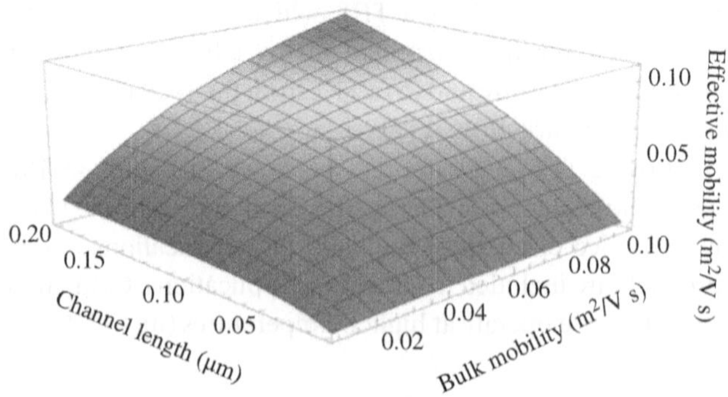

Figure 3.9 Dependencies of the effective field-effect mobility on the gate length for different bulk electron mobilities (for nondegenerate 2DEG, see Eq. (3.4)). Source: Adapted from Shur et al. [16].

result, we have to account for the effects of the electron inertia and for the quasi-ballistic transport invoking the concept of ballistic mobility [13–16], dominating when the channel scattering rate due to the electronic collisions at the channel boundaries.

Figure 3.9 shows the dependencies of the effective field-effect mobility on the gate length for different bulk electron mobilities μ. These dependencies account for ballistic mobility,

$$\frac{1}{\mu} = \frac{1}{\mu_{bulk}} + \frac{1}{\mu_{bal}} \tag{3.3}$$

For the nondegenerate two-dimensional electron gas (2DEG)

$$\mu_{bal} = \frac{L}{2v_{th}} = \frac{Lm_{eff}^{1/2}}{(2\pi k_B T)^{1/2}} \tag{3.4}$$

Here, v_{th} is the thermal velocity, m_{eff} is the electron effective mass, k_B is the Boltzmann constant, and T is temperature. As seen from Eqs. (3.3) and (3.4) and Figure 3.10, at an ultra-short channel length, L, the advantage of high bulk electron mobility μ_{balk} diminishes and, hence, TFTs having smaller mobility than silicon are at less of a disadvantage. Figure 3.9 shows that at 20 nm channel length the expected advantage of having a high mobility material in a FET channel is marginal.

Figure 3.10 compares the calculated ballistic mobility for the nondegenerate two-dimensional electron gas (2DEG) for the effective electron mass of ZnO $m_{eff} = 0.28$ [17] and the electron mobility of ZnO of 345 $cm^2/V\,s$ [11] with that for Si NMOS assuming the electron mobility of 650 $cm^2/V\,s$ and the effective mass of 0.19. As seen, for very short channel lengths, these dependencies come closer.

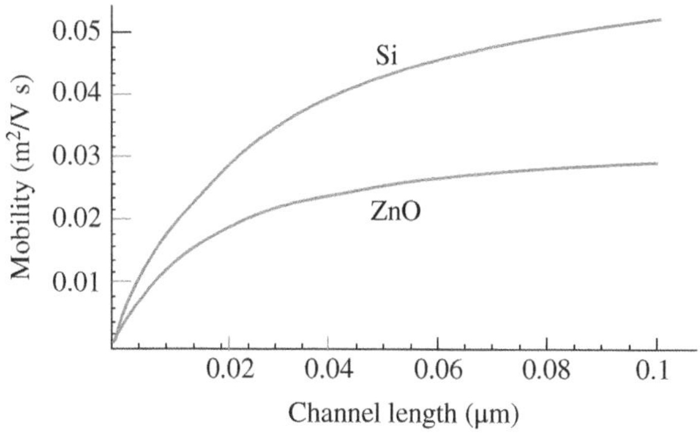

Figure 3.10 Ballistic mobility for the nondegenerate 2DEG for ZnO of and Si NMOS.

Figure 3.11 Transparent IGZO TFTs on polyethylene terephthalate (PET) substrate. Source: Adapted from Mukherjee et al. [18].

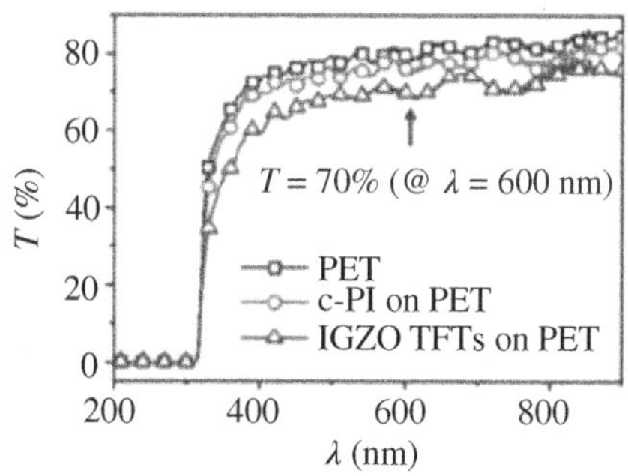

Figure 3.11 shows that IGZO TFTs have excellent transparency (over 98% in the most transparent devices in the 350–850 nm, i.e., in the entire visible range and beyond) [18].

Table 3.1 compares a-Si, poly-Si, and oxide TFT technologies. As seen, oxide TFTs have many advantages compared to more mature leading a-Si and poly-Si TFT technologies and have the potential for a large variety of applications. The most important advantages of oxide TFTs are low cost, high reliability, high speed (linked to high mobility), high on-current, and scalability. These advantages are crucial for display applications. Oxide TFTs enable larger area displays with reduced flicker (due to a higher speed of operation) with increased pixel density, resulting in sharper and more detailed images and improved color quality. A low processing temperature makes oxide TFT compatible with a wider range of substrate materials, including flexible plastics. A high uniformity of oxide TFTs over large areas allows for maintaining consistent performance across a range of operating conditions.

Table 3.1 Qualitative comparison of characteristics of a-Si, poly-Si, and oxide TFTs.

	a-Si TFT	Poly-Si TFT	Oxide TFT
TFT channel	n-channel	n- and p-channels	n- and p-channels
Mobility ($cm^2/V\,s$)	0.5–1	100	10–300
On current	Low	High	High
Power consumption	Low	High	Low
Off current	Medium	Medium	Low
Number of masks	4	5	5
Cost	Low	Medium	Low
Process temperature	150–350	250–550	25–400
Temperature stability	Poor	Good	Good
Reliability	Medium	Good	Good
Scalability	Good	Medium	Good
Uniformity	Medium	Good	Good
Manufacturing cost	Low	High	Low

3.3 N-type, P-type, and Ambipolar Oxide TFTs

Table 3.2 lists the electrical properties of oxides used for n-channel, p-channel, and ambipolar TFTs [18, 39]. The performance of p-channel oxide TFTs is inferior to that of the n-channel device, e.g., smaller mobilities and worse stability [38]. Record values of the electron mobility in oxide TFT could exceed 100 $cm^2/V\,s$. For short channel devices, these mobility values are sufficient even for RF applications. Most of the oxide TFTs support transparent display technology and have a high on-to-off ratio. In particular, In_2O_3 oxide TFTs are compatible with the hydrogen-rich CMOS BEOL process making these devices suitable for 3D integration applications [40].

For complementary (CMOS) operation, p-type polysilicon [40], crystalline Si PMOS, organic p-type TFTs, or p-type carbon nanotubes [40–43] could be used in combination with n-channel oxide TFTs. However, organic p-type TFTs might have stability problems and lower mobilities, typically below 2 $cm^2/V\,s$. Further development of higher performance p-type oxide TFTs could enable better ultra-low-power complementary TFT circuits for transparent displays.

Both normally off and normally on oxide TFTs have been fabricated with threshold voltages of −7.3 V for In_2O_3/ZnO TFTs [44], 3.1 V for In_2O_3

Table 3.2 Oxide TFT properties related to channel materials.

Material	Channel type	Bandgap (eV)	Mobility ($cm^2/V\,s$)	I_{on}/I_{off}	Trans-parency	References
IGZO	n-channel	3.5	10–40, 75 (lab)	10^5–10^{10}	Yes	[19–22]
InGaO	n-channel	3.6–4.7	10.8–12.9	—	Yes	[23]
ZnO[a)]	n-channel p-channel	3.3	150 (crystalline) 4–85 (TFT)	10^7–10^9	Yes	[24–28]
In_2O_3	n-channel	2.61	91.3	10^7	Yes	[29, 30]
SnO	p-channel Ambipolar	0.5 2.27 (direct)	7–10	10^5	No	[31]
CuO[b)]	p-channel	>2.1	100 (crystalline) 0.4 (TFT)	—	Yes	[31–33]
NiO[c)]	p-channel	3.7	1.6×10^{-4}	130	Yes	[34]
CuI	p-channel	~3	40 (calculated) 5 (TFT)	10^7	Yes	[35]

a) High breakdown voltage.
b) A relatively unstable material, which can lead to TFT lifetime problem.
c) A very stable material.

TFTs [45], 1.1 V for ZnO TFTs [46], 2.5 V for SnO TFTs [47], and −0.5 V for IGZO TFTs [48].

The field effect mobility values of n-channel oxide TFTs vary depending on the fabrication process. The highest ZnO TFT mobility was 345 $cm^2/V\,s$ [11]. For spray pyrolyzed ZnO film the field effect mobility of 41.53 $cm^2/V\,s$ was achieved [49]. ZnO TFTs fabricated by radio frequency magnetron sputtering achieved mobility values up to 70 $cm^2/V\,s$ [50]. Amorphous Indium-gallium-zinc-oxide (a-IGZO) TFTs have mobility values up to 70 $cm^2/V\,s$ comparable to ALD deposited IGZO(~74 $cm^2/V\,s$) [51]. In_2O_3 TFTs have mobility values up to 139 $cm^2/V\,s$ [52]. The mobility of an oxide TFT is dependent on the composition material and fabrication process, which are discussed in Chapters 2 and 4 of this book.

Figures 3.12 and 3.13 show the subthreshold slope (*SS*) and on-to-off current ratio (I_{on}/I_{off}), respectively. Data for IGZO, ZnO, In_2O_3, and SnO are from Refs. [53–56].

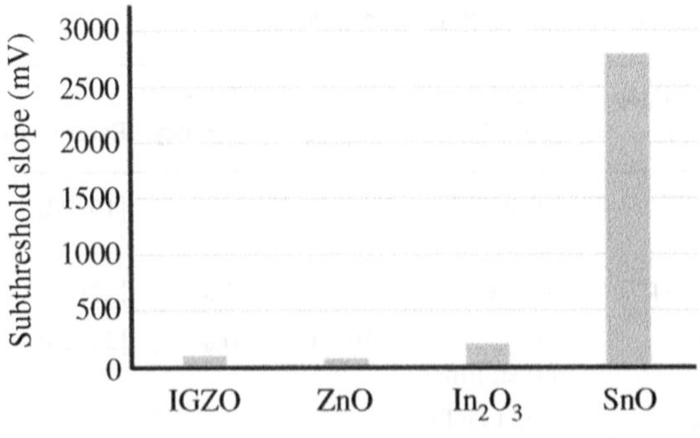

Figure 3.12 Subthreshold slope for different oxide TFTs.

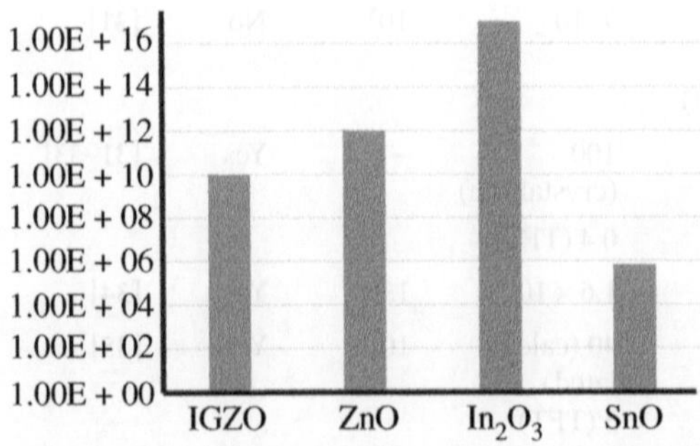

Figure 3.13 I_{on}/I_{off} ratio for different oxide TFTs.

Ambipolar oxide TFTs [57] can allow more compact CMOS architectures and be used in reconfigurable artificial synapses for simulating the biological nervous system [58]. However, high bandgaps and states in the bandgap of metal oxides inhibit ambipolar behaviors. SnO is more suited for the ambipolar behavior due to its low electron effective mass, low hole effective mass, and small indirect bandgap.

Metal oxide heterojunction TFTs, such as using In_2O_3/ZnO heterojunction channels [59], have demonstrated high electron mobility of 19 cm^2/V s, I_{on}/I_{off} of $>10^6$, and SS of 108 mV/dec. They can be manufactured on polymer substrates [60].

3.4 Oxide TFT Designs

Just as a-Si or poly-Si TFTs, oxide TFTs use top-gate and bottom-gate designs, depending on the position of the gate electrode, as shown in Figure 3.14a–d. The gate contact could be coplanar or staggered with respect to the source and drain contacts for both configurations. In staggered

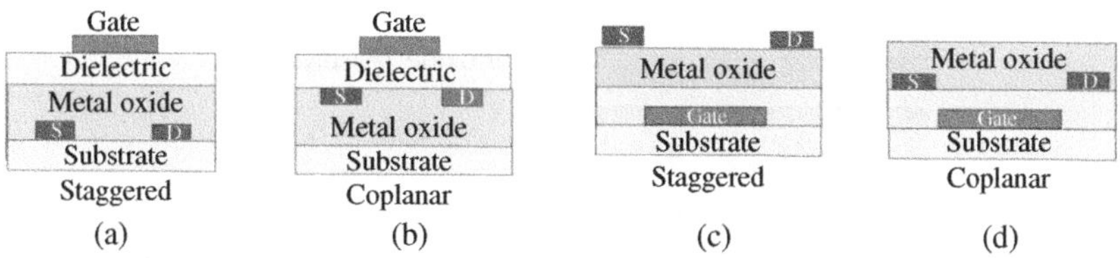

Figure 3.14 Schematic designs of oxide TFTs with (a) top-gate staggered, (b) top-gate coplanar, (c) bottom-gate staggered, (d) bottom-gate coplanar structures.

structures, the current flows vertically to the channel and horizontally from source to drain. Transparent top-gate oxide TFT designs could also use ITO contacts [61].

The bottom gate designs for oxide TFTs exhibit better control over threshold voltage variations, have a lower off-state leakage current, and better device stability. This design is suitable for applications in LCDs because the light from the backlight unit is shielded by the metal gate electrode. The bottom gate oxide TFTs integrated with a sensing pad or incorporating microfluidic channels have also been used as biosensors [62–64].

The top-gated structures could avoid overlap between the gate and the source/drain contacts, and, consequently, have a smaller parasitic capacitance. It also allows the processing of a semiconductor at higher temperatures without damaging previously deposited layers. In the coplanar oxide TFTs, the channel directly contacts the source and drain and the current flows horizontally in the channel.

Emerging novel designs, such as ambipolar, folded-gate complementary, vertical, and double-gated oxide TFTs, have been reported [65], as shown in Figure 3.15a–d. In another example, a design with a suspended gate with an air gap as the insulator layer has also been used to measure the pressure [66]. Special TFT structures used for various sensor applications are discussed in Chapter 5 Applications.

Figure 3.16 shows typical material choices for oxide TFT layers. Additionally, metal oxide heterojunctions need to be further explored [67].

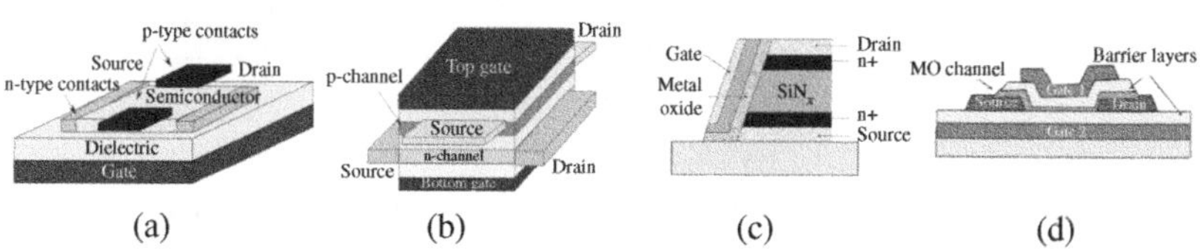

Figure 3.15 (a) Ambipolar, (b) folded-gate complementary, (c) vertical, and (d) double-gated oxide TFTs.

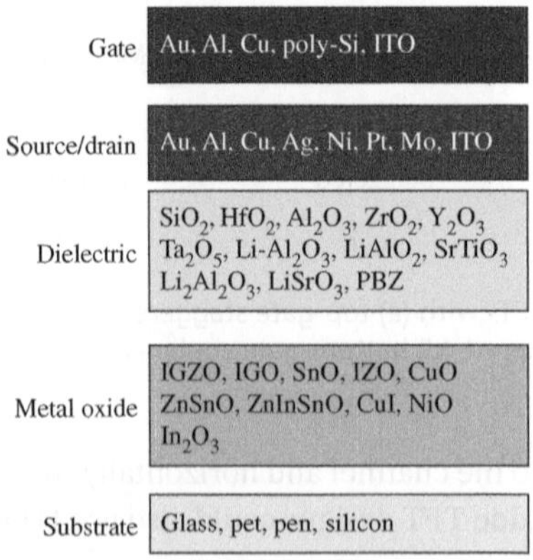

Figure 3.16 Typical material choices for oxide TFT layers.

3.5 Modeling of Oxide TFTs

Oxide TFTs are modeled to predict their electrical behavior using compact models [68, 69], implemented in circuit simulators, such as advanced device system (ADS) [70] or PSPICE [71] and TCAD simulators [72, 73]. These models have to account for material properties of the TFT constituent layers and for parasitic elements, such as contact resistances and parasitic capacitances. These two approaches could be combined, as shown in Figure 3.17 [74, 75]. Modeling is the key component of device technology co-development (DTCO). Figure 3.18 illustrates the integrated modeling approach, which starts from *ab initio* calculations of material properties, such as band diagrams, effective masses, velocity-field characteristics, polarization, elastic and piezoelectric tensors, and defect properties, including the development of dislocation arrays. These parameters are fed into TCAD software and SPICE software for the simulation of device structures, and

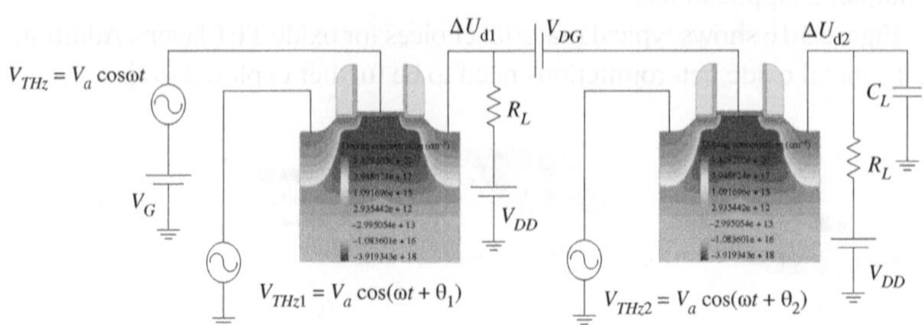

Figure 3.17 TCAD mixed mode: SPICE simulation with device represented by the hydrodynamic TCAD model [74, 75].

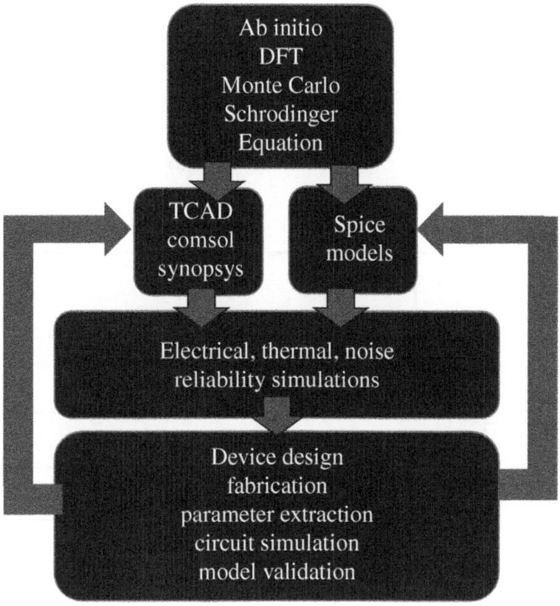

Figure 3.18 DTCO approach.

their electrical, thermal, and even reliability properties. These simulations are accompanied by device characterization and model validation steps in a feedback loop. Artificial intelligence (AI) models could be very effective in assisting this DTCO process.

Figure 3.19a shows the equivalent circuit of a TFT that accounts for parasitic and leakage elements and the electron inertia represented by the Drude inductance [76]. The multisegmented approach represented by Figure 3.19b accounts for the distribution of the carrier density and current along the channel [77]. N is the number of segments.

For short channel oxide TFTs, the contact impedances might play an important or even dominant role, and an oxide TFT compact model should account for contact impedance nonlinearity, as shown in Figure 3.20 [78]. In ultra-short channel devices, the intrinsic (bulk) mobility of a TFT active layer plays a diminished role and simulations show that oxide TFTs could respond to and detect impinging radiation at sub-THz and THz frequencies and therefore are capable of supporting applications in future 6G communication systems. To adequately simulate the response at such high frequencies requires splitting the channel into sections to account for the nonuniform distribution of AC current and carrier concentration across the channel.

A new compact model developed for oxide TFTs was used to analyze the response of oxide TFTs in signal detection circuits [79]. This improved

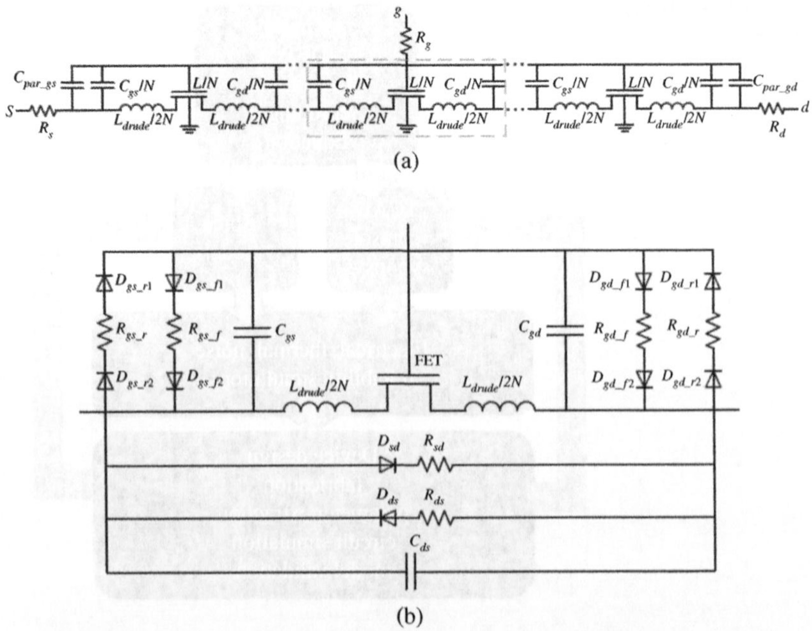

Figure 3.19 (a) Equivalent circuit of a multi-segment SPICE model for TFT. Source: Liu et al. [76]/MDPI/CC-BY-4.0, (b) equivalent circuit for each segment including leakage components. Source: Adapted from Liu et al. [76]/MDPI/CC-BY-4.0.

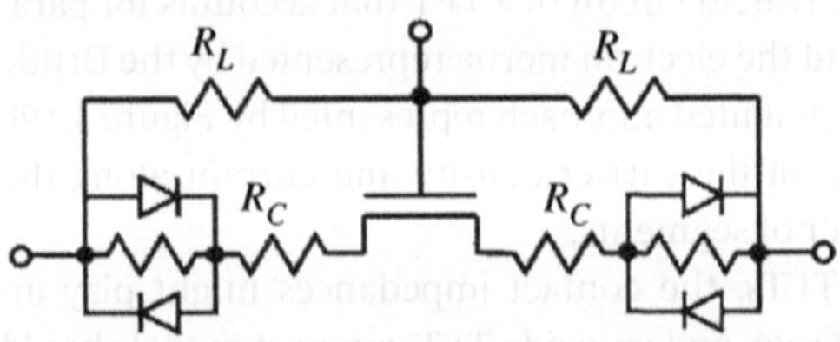

Figure 3.20 Compact TFT equivalent circuit with contacts. Source: Gundlach et al. [78]/AIP Publishing.

compact model for oxide TFTs combines the RPI TFT model, the unified charge control model (UCCM), and the multi-segment TFT compact model. It accounts for non-exponential subthreshold slopes and nontrivial capacitance dependence on gate bias, which are not captured by conventional TFT models. A non-exponential slope in the subthreshold regime was accounted for by introducing a varying subthreshold slope in contrast to conventional MOSFET models that use a constant subthreshold slope.

The Drude inductances included in the SPICE model

$$L_s = L_d = L_{drude}/2 \tag{3.5}$$

account for electron inertia [80].

$$L_{drude} = \tau R_{ch} \tag{3.6}$$

Here, R_{ch} is the channel resistance, $\tau = m\mu/q$ is electron momentum relaxation time, and m is the electron effective mass. This could be understood from the famous Drude equation

$$Z = R(1 + i\omega\tau) = R + iR\omega\tau = R + i\omega L_{drude} \tag{3.7}$$

Here, R is a DC impedance. As seen from Eqs. (3.6) and (3.7) scattering occurs on the time scale on the order of the momentum relaxation time τ.

The measured effective electron mass in ZnO is 0.28. Using the highest ZnO TFT mobility of 345 cm^2/V s, we find τ = 55 fs. The calculated mean free path at room temperature is 12 nm. Therefore, electrons in a 20 nm ZnO channel will behave as a quasi-ballistic electronic fluid [81]. Oscillations (waves) of the electronic density of such a fluid are called plasma waves. These waves are the consequence of electron inertia. An important parameter characterizing the high-frequency response of such a device is the fundamental plasma frequency, f_p, of the two-dimensional electron gas in the TFT channel given by

$$f_p = \frac{s}{4L} \tag{3.8}$$

The plasma velocity, s, is the function of the gate voltage swing $V_{gt} = V_{gs} - V_T$, where V_{gs} is the gate-to-source voltage, V_T is the threshold voltage, and L is the channel length.

$$s = v_{th}\sqrt{\left(1 + \exp - \frac{qV_{gt}}{\eta k_B T}\right) \ln\left(1 + \exp - \frac{qV_{gt}}{\eta k_B T}\right)} \tag{3.9}$$

Here, $v_{th} = \left(\frac{3k_BT}{2m_{eff}}\right)^{1/2}$ is the thermal velocity, m_{eff} is effective mass, k_B is the Boltzmann constant, and T is temperature. For IGZO, $m_{eff} \approx 0.16$ [82].

Figure 3.21 shows the calculated plasma frequency and the quality factor

$$Q = 2\pi f_p \tau = \omega_p \tau \tag{3.10}$$

These calculations show that a ZnO TFT could operate as a sub-THz or THz detector. When the quality factor is much less than unity, the detection mechanism is rectification of the THz voltage, V_{THz}, induced by impinging radiation.

Due to the nonlinearity of the oxide TFT current-voltage characteristic the AC current induced by the THz voltage is given by

$$\begin{aligned} i_{AC} &= aV_{THzo}\,\mathrm{Cos}\omega t + bV_{THzo}^2\,\mathrm{Cos}^2\omega t \\ &= aV_{THzo}\,\mathrm{Cos}\omega t + \frac{1}{2}bV_{THzo}^2\,\mathrm{Cos}2\omega t + \frac{1}{2}bV_{THzo}^2 \end{aligned} \tag{3.11}$$

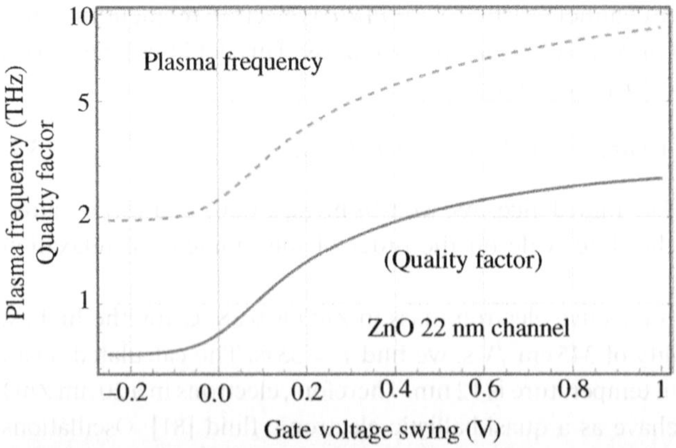

Figure 3.21 Calculated plasma frequency and quality factor for 22 nm ZnO TFT. Parameters used in the simulation: barrier layer dielectric constant 16, barrier thickness 8 nm, mobility 345 cm^2/V s, $m_{eff} = 0.28$, ideality factor 1.8.

Equation (3.11) is the series expansion of the nonlinear response keeping the second-order term. The last term in this expansion is the DC response

$$V_r = \frac{1}{2} b V_{THzo}^2 \tag{3.12}$$

When the quality factor is larger than unity the device responds as a resonant nonlinear transmission line, as shown in Figure 3.19a, and plasma waves (the waves of electron density oscillations) propagate in the oxide TFT channel as schematically shown in Figure 3.22a. When quality factors are smaller than unity, the plasma waves decay as shown in Figure 3.22b.

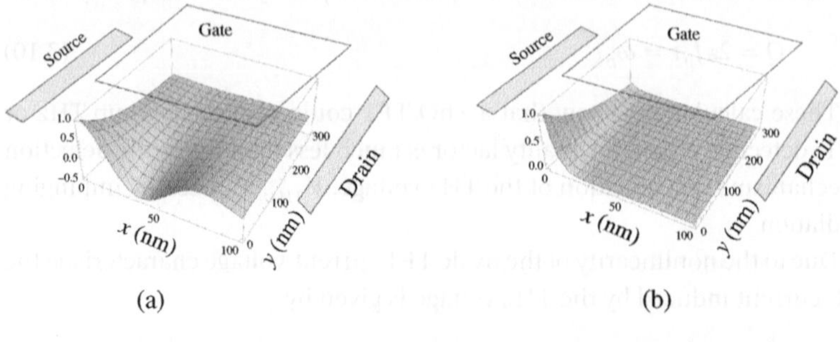

Figure 3.22 Oscillations of electron density (arbitrary units) in an oxide TFT channel (plasma waves) excited by impinging radiation: (a) resonant plasma waves, (b) damped waves.

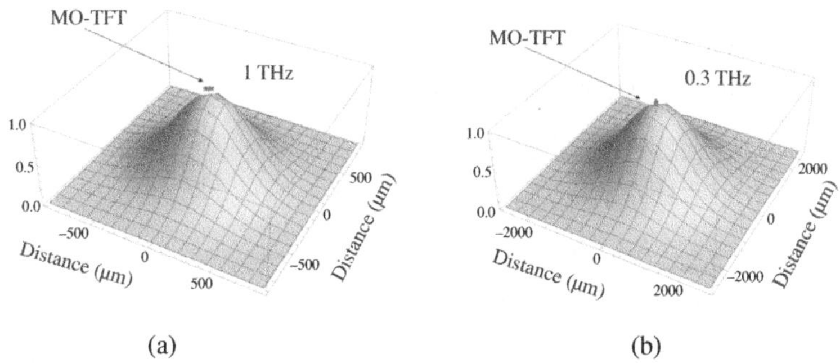

Figure 3.23 Intensity distributions of focused radiation at (a) 1 THz and (b) 0.3 THz and dimensions of oxide TFTs.

The rectified voltage between the source and drain can be expressed by the following equation [83]:

$$V_r = \frac{U_a^2}{4V_{gte}}\left(1+\beta-\frac{1+\beta\,\mathrm{Cos}(2k_1L)}{\mathrm{Sinh}^2(k_2L)+\mathrm{Cos}^2(k_1L)}\right) \tag{3.13}$$

Here,

$$\beta = \frac{2\omega\tau}{\sqrt{1+\omega^2\tau^2}};\ k_1 = \frac{\omega}{s}\sqrt{\frac{(1+\omega^{-2}\tau^{-2})^{1/2}+1}{2}};$$

$$k_2 = \frac{\omega}{s}\sqrt{\frac{(1+\omega^{-2}\tau^{-2})^{1/2}-1}{2}} \tag{3.14}$$

The wavelength of the THz radiation is much larger than typical oxide TFT dimensions and, therefore, the response spatial resolution is defined by the dimension of the oxide TFT channel, as shown in Figure 3.23 [84, 85]. These figures demonstrate the advantages of using TFT arrays for high-frequency applications because such arrays could capture a much larger fraction of the impinging THz beam. Using asymmetrical unit cell structures greatly increases the efficiency and responsivity [86].

Figure 3.24 shows the equivalent circuit for registering oxide TFT THz response [76].

Figure 3.24 Equivalent circuit for registering oxide TFT THz response. Source: Shur et al. [79]/MDPI/CC-BY-4.0.

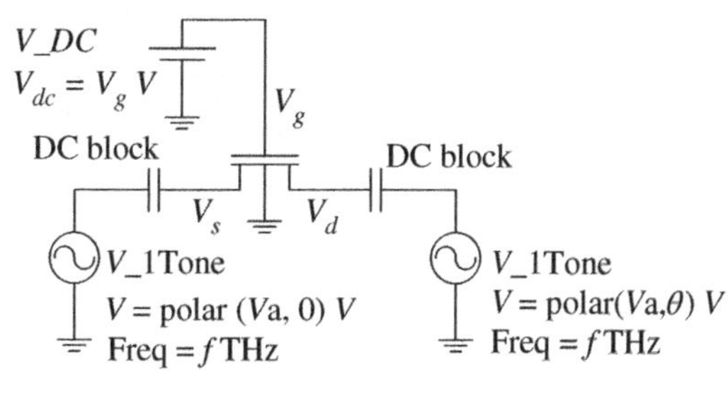

The improved TFT compact model is the extension of the TFT SPICE UCCM model described in references [87, 88] and implemented in Verilog-A. The basic equations of the model are:

$$I_{ds} = \frac{g_{chi} V_{ds}(1 + \lambda V_{ds})}{\left[1 + (V_{ds}/V_{sate})^m\right]^{1/m}} \tag{3.15}$$

$$I_{sat} = \frac{g_{chi} V_{gte}}{1 + \sqrt{1 + (V_{gte}/V_L)^2}} \tag{3.16}$$

$$g_{chi} = q n_s \mu W/L \tag{3.17}$$

$$V_{sate} = \frac{I_{sat}}{g_{chi}} \tag{3.18}$$

$$V_{gte} = \eta V_{th}\left[1 + \frac{V_{gt}}{2\eta V_{th}} + \sqrt{\delta^2 + \left(\frac{V_{gt}}{2\eta V_{th}} - 1\right)^2}\right] \tag{3.19}$$

$$V_L = \frac{v_s L}{\mu} \tag{3.20}$$

$$\mu = \mu_0 \frac{\alpha V_{gte}{}^{\gamma}}{1 + \alpha V_{gte}{}^{\gamma}} \tag{3.21}$$

$$\eta = \eta_2 + \frac{\eta_2 - \eta_1}{1 + \exp\left(\frac{V_{gst} - V_{gs}}{\Delta V_{gst}}\right)} \tag{3.22}$$

$$C_{gs} = \frac{2}{3} C_{gc}\left[1 - \left(\frac{V_{gte} - V_{dse}}{2V_{gte} - V_{dse}}\right)^2\right] \tag{3.23}$$

$$C_{gc} = \frac{\varepsilon_i LW \exp(V_{gt}/(\eta_c V_{th}))}{(d_i + \alpha_g V_{gt})(1 + \exp(V_{gt}/(\eta_c V_{th})))} \tag{3.24}$$

$$C_{gd} = \frac{2}{3} C_{gc}\left[1 - \left(\frac{V_{gte}}{2V_{gte} - V_{dse}}\right)^2\right] \tag{3.25}$$

$$V_{dse} = \frac{1}{2}\left[V_{ds} + V_{gte} - \sqrt{(\eta_c V_{th} \delta_c)^2 + (V_{ds} - V_{gte})^2}\right] \tag{3.26}$$

$$\eta_c = \eta(1 + \beta V_{gte}) \tag{3.27}$$

$$n_{s0} = n_0 \ln(1 + \exp(V_{gt}/(\eta V_{th}))) \tag{3.28}$$

$$n_0 = \varepsilon_i \eta V_{th} / (q(d_i + \alpha_g V_{gt})) \tag{3.29}$$

$$V_{gt} = V_{gs} - V_T + \sigma V_{ds} \tag{3.30}$$

$$\lambda_e = \lambda - \frac{\lambda_g}{1 + \exp\left(\frac{V_{go} - V_{gs}}{V_{\lambda g}}\right)} - \frac{\lambda_d}{1 + \exp\left(\frac{V_{do} - V_{ds}}{V_{\lambda d}}\right)} \tag{3.31}$$

Table 3.3 defines the parameters of the oxide TFT model.

The key physics of a TFT in the above compact model is captured by the UCCM. The UCCM equation (Eq. (3.28)) describes the below and above threshold regimes of operation controlled by the gate voltage swing U_o. An improved compact model based on the RPI TFT model accounts for a

Table 3.3 Parameters used in TFT models.

Parameters	Notation	Unit
Drain-to-source and saturation currents	I_{ds}, I_{sat}	A
Extrinsic drain-to-source and gate-source voltages	V_{dse}, V_{gs}	V
Knee shape of the output characteristics parameter	m	—
Intrinsic linear channel conductance	g_{chi}	S
Gate voltage swing and effective swing	V_{gt}, V_{gte}	V
Gate voltage-dependent and low field mobilities	μ, μ_0	$m^2/(V\,s)$
Gate voltage mobility dependence parameters	α and γ	
Subthreshold, moderate, deep subthreshold parameters	η, η_1, η_2	
Voltage and transition separating two subthresholds	V_{gst}. ΔV_{gst}	V
Gate width, length, barrier thickness	W, L, d_i	m
Parameter describing the transition to saturation in n_s	κ	
2D electron sheet densities	n_s, n_{so}, n_o	m^{-2}
Dielectric layer permittivity	ε_i	F/m
Gate voltage n_s dependence and DIBL parameters	α_g, σ	
Threshold, velocity, and effective saturation voltages	V_T, V_L, V_{sate}	V
Saturation velocity	v_s	m/s
Above to subthreshold region transition parameter	δ	
Differential, gate-to-drain, gate-to-source capacitance	C_{gc}, C_{gd}, C_{gs}	F
Capacitance subthreshold, ideality, transition	η_c, β, δ_c	
Gate length modulation parameters	λ, λ_e, λ_g, λ_d	1/V
Gate and drain voltage for gate length modulation	$V_{\lambda g}$, $V_{\lambda d}$	V

non-exponential slope in the subthreshold regime by introducing a varying subthreshold slope, as shown in Eq. (3.22), and a nontrivial capacitance dependence on the gate bias. These new features allowed us to obtain an excellent agreement with the measured I–V/C–V characteristics for both long and short n-channel and p-channel TFTs.

Parameter λ_e describes the output conductance by accounting for the output conductance dependence on both drain-to-source and gate-to-source voltages.

Figures 3.25–3.27 present the results of comparing the experimental data [85] with simulations using models described in Figure 3.19 on oxide

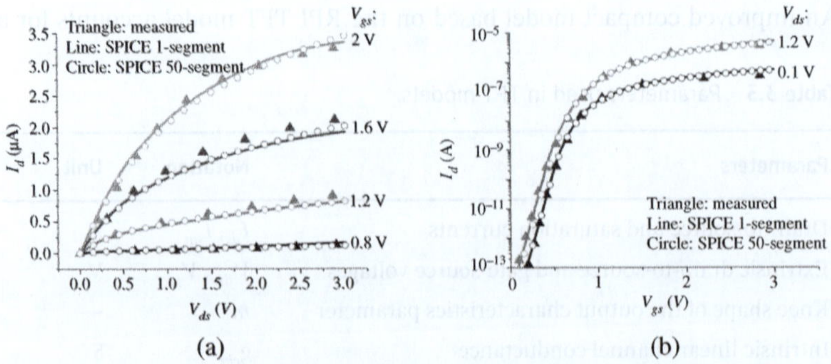

Figure 3.25 (a) Output characteristics and (b) transfer characteristics of a 27-nm trench-gate-self-aligned IGZO TFT. "▲" measured data, "-" one-segment SPICE model, "o" 50-segment SPICE model. Source: Adapted from Matsubayashi et al. [85].

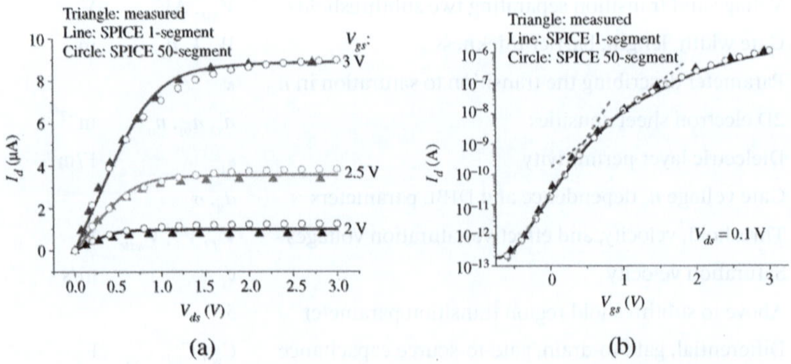

Figure 3.26 (a) Output characteristics and (b) transfer characteristics of a 20 μm n-channel oxide TFT. "▲" measured data, "-" one-segment SPICE model, "o" 50-segment SPICE model. Dashed lines in (b) illustrate the parameter extraction for the two subthreshold slopes [79]. Source: Adapted from Matsubayashi et al. [85].

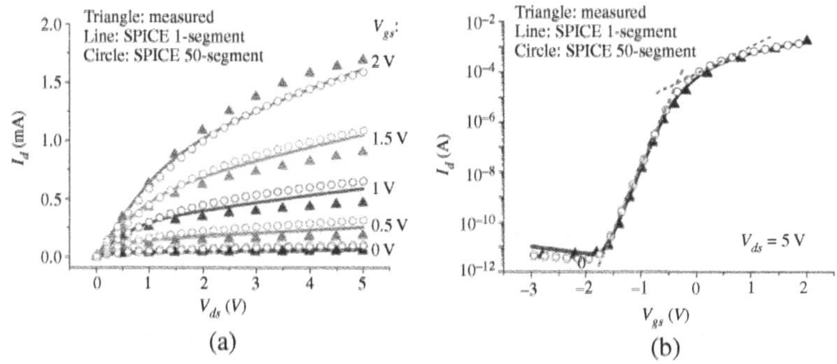

Figure 3.27 (a) Output characteristics and (b) transfer characteristics of a 0.8 μm n-channel oxide TFT. "▲" measured data, "-" one-segment SPICE model, "o" 50-segment SPICE model. Dashed lines in (b) illustrate the parameter extraction for the two subthreshold slopes [76, 79].

TFTs of a wide range of channel lengths. Figure 3.25 shows the (a) output characteristics and (b) transfer characteristics on a 27-nm trench-gate-self-aligned IGZO TFT from experimental measurement as well as from 1-segment and 50-segment SPICE model simulations. As seen, both 1-segment and 50-segment simulations agree well with the measured data. The multisegmented model is required for the simulation of high-frequency TFT response.

Figure 3.26 shows the (a) output characteristics and (b) transfer characteristics on a 20 μm n-channel oxide TFT from experimental measurement as well as from 1-segment and 50-segment SPICE model simulations.

Figure 3.27 shows the (a) output characteristics and (b) transfer characteristics on a 0.8 μm n-channel oxide TFT from experimental measurement as well as from 1-segment and 50-segment SPICE model simulations.

Table 3.4 presents the parameters used in the above simulations. These parameters are provided in references containing experimental data, except the capacitances that were calculated using the gate dielectric thickness and dielectric constant but then adjusted to fit the measured data.

Table 3.4 Parameters used in the simulation. Effective mass 1, saturation velocity 10^5 m/s.

Material	*L/W* (μm)	C_{gate} (F/m)	Mobility (m^2/V s)	Threshold voltage (V)
IGZO	0.027/0.032	0.0031	0.00065	0.52
IGZO	0.8/250	0.00069	0.0008	−0.5
IGZO	20/20	0.0046	0.0012	1.1

Figure 3.28 shows the effect of subdividing the TFT channel into segments of the spatial distributions of the channel parameters. The simulated response using the 1-segment and 50-segment models to the impinging THz radiation is quite different. The comparison with the Si MOSFET response shows that the IGZO TFT has a smaller but comparable response because the difference in the field-effect mobility does not dramatically change the response for such short-channel transistors [76].

As seen in Figure 3.29, the difference between the 5- and 50-segment models is especially important near the drain (as expected). The difference is the largest for the gate-to-drain capacitance. This difference affects the predicted response to THz radiation. Figure 3.29 compares simulated TFT and Si MOSFET responses using 1-segment and 50-segment SPICE models [79]. For both oxide TFT and n-MOSFET, the difference between 1-segment and 5-segment simulations is very significant even though the maximum response values for both models are similar.

Figure 3.30 shows an equivalent circuit of an oxide TFT accounting for parasitic capacitances [79]. C_{frs}, C_{frd}, and C_{frg} are gate-to-source, gate-to-drain, and drain-to-ground fringing capacitances, respectively; Z_a and Z_L are the source internal impedance and the load impedance, respectively. This circuit might be important at sub-THz and THz frequencies because their significance is proportional to the d_i/L ratio where d_i is the barrier layer thickness and L is the gate length. This ratio tends to be large in short-channel TFTs. Also, the feedback conductance connecting the drain terminal (output) and the gate terminal (input) is proportional to $2\pi f C_{frd}$ where f is frequency strongly affects the cutoff frequency and maximum frequency of operation [89]. At very high frequencies, the source impedance could be an antenna capturing impinging radiation.

The transistor symbol corresponds to the equivalent circuit shown in Figure 3.19b [77]. This equivalent circuit accounts for leakage currents via the introduction of multiple diodes representing different leakage channels. It also accounts for the Drude inductances, L_s and L_d, representing the effects of electron inertia. Such a model could be used for evaluation of the effect of parasitic elements on the device performance. It could help identify which parasitic elements play a crucial role. The inclusion of the Drude inductances (describing the effect of the electron inertia) makes the model suitable for high-frequency simulations.

Figure 3.31 shows the calculated C–V characteristics in comparison with the estimated parasitic gate-to-source and gate-to-drain capacitances, which are estimated as

$$C_{gsp} = C_{gdp} = \alpha \varepsilon \varepsilon_o W/2 \tag{3.32}$$

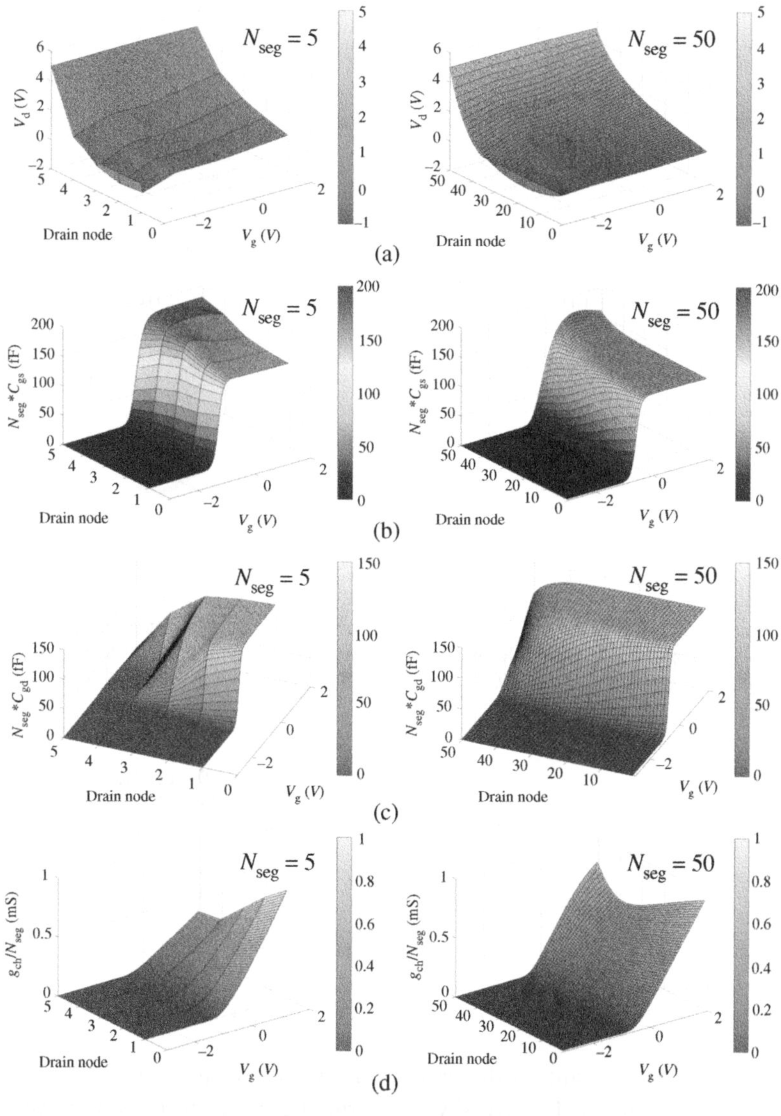

Figure 3.28 Profiles of the distributed parameters for the TFT-UCCM compact model with 5 segments (Nseg = 5) and 50 segments (Nseg = 50): (a) potential at the drain node of each segment in the channel, (b) normalized gate-to-source capacitance for each segment, (c) normalized gate-to-drain capacitance for each segment, (d) normalized channel conductance for each segment. Source: Liu et al. [76]/MDPI/CC-BY-4.0.

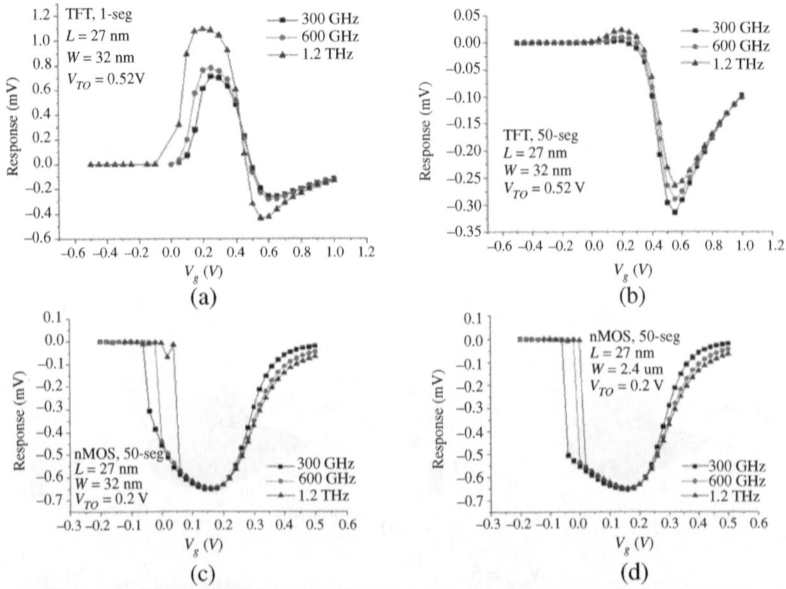

Figure 3.29 Simulated TFT with V_{th} 0.52 V responses using (a) 1-segment and (b) 50-segment SPICE models, and simulated n-MOSFET with V_{th} 0.2 V responses using (c) 1-segment and (d) 50-segment SPICE models. Source: Adapted from Shur et al. [79].

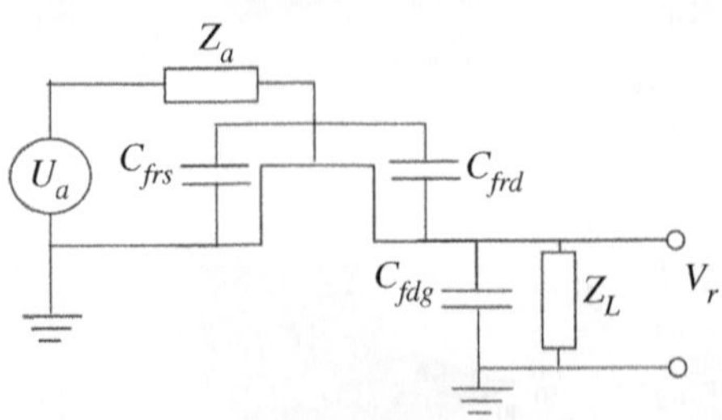

Figure 3.30 Equivalent circuit of an oxide TFT accounting for parasitic capacitances.

Here, α is constant on the order of unity and taken as unity. This is a very crude estimate by necessity, since fringing capacitance is a very weak (logarithmic) function of distance and, therefore would depend on specific layouts.

As seen from Figure 3.31, the values of C_{gs} calculated for the 50-segment model are very close to those calculated for the conventional 1-segment model. However, this difference is quite important for C_{gd} because the

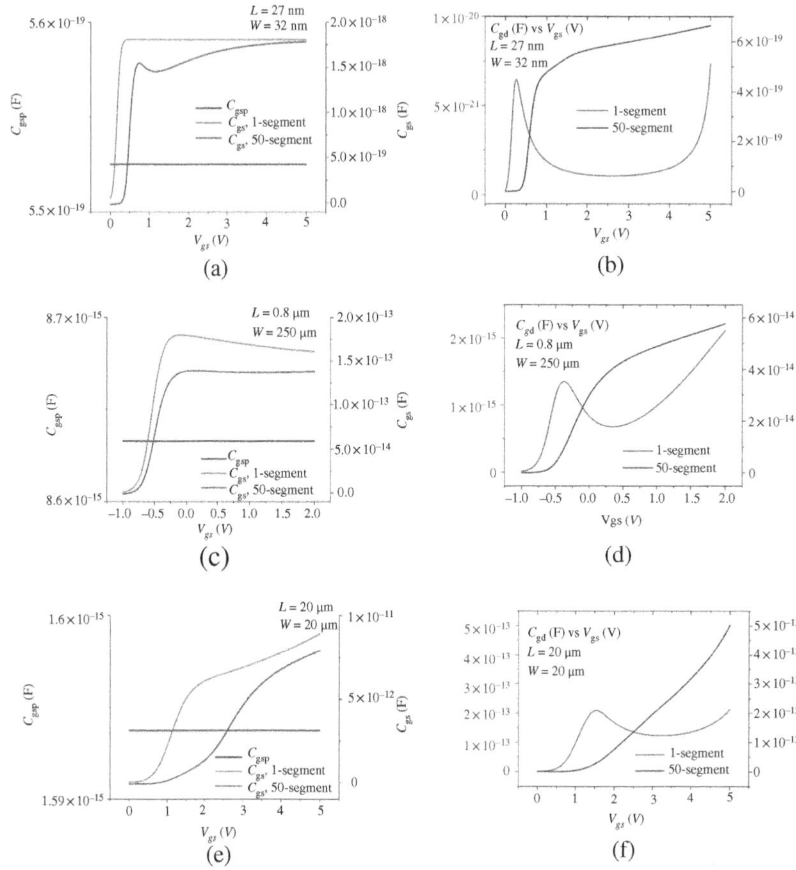

Figure 3.31 Calculated *C-V* characteristics at $V_{ds} = 5$ V in comparison with the estimated parasitic gate-to-source and gate-to-drain capacitances C_{gsp}. (a,b) 27 nm channel; (c,d) 0.8 μm channel; (e,f) 20 μm channel. Source: Shur et al. [79]/IOP Publishing/CC BY 4.0.

segmentation accounts much more accurately for the channel nonuniformity at the drain bias of 5 V used in the calculation. Also, parasitic capacitances are quite important being comparable to the channel capacitances for ultra-short channel devices.

A THz TFT detector could be implemented as an inverter. Figure 3.32 shows THz detector schematics and response to 300 and 600 GHz radiation computed using a 1-segment TFT SPICE model. As seen, the largest

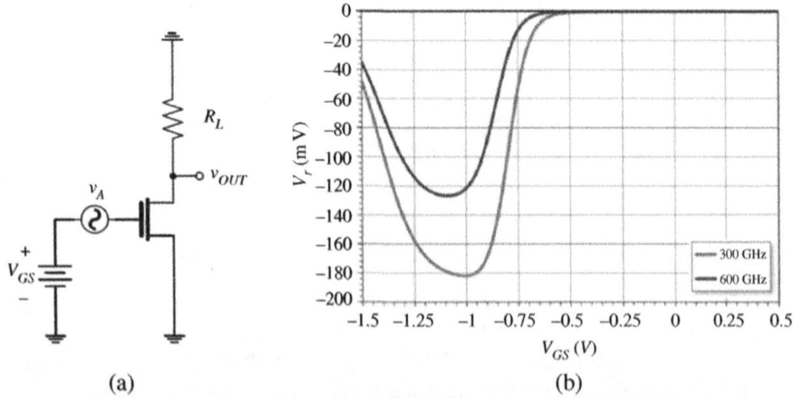

Figure 3.32 (a) A THz detector schematic diagram and (b) the response with 1-segment TFT SPICE model. Channel length and width $L = 0.8\,\mu m$, $W = 250\,\mu m$, $V_{th} = -0.5$ V, parasitic capacitances $C_{frs} = C_{frd} = C_{fdg} = 2.7$ fF, THz amplitude: 10 mV. The THz source has a source resistance of 50 Ω. Source: Shur et al. [79]/IOP Publishing/CC BY 4.0.

response is below the threshold, where the transistor current-voltage characteristic is strongly nonlinear (see Figure 3.31). The rectification of THz radiation by a 0.8 mm channel yields a fairly large frequency-dependent response. Such frequency dependence could be used to calibrate an oxide TFT as a THz spectrometer.

Figure 3.33 shows (a) the equivalent circuits of a complementary oxide TFT inverter and (b) its response to impinging sub-THz radiation [79]. The response magnitude is much smaller than for an n-channel inverter, as shown in Figure 3.32, but it is still sufficient for registering the intensity of impinging THz radiation.

Oxide TFTs could also be used as detectors sensing both the intensity and phase of impinging THz radiation [75, 88, 90–96]. Figure 3.34 shows a FET driven by the signals from two antennas with the signals shifted by phase angle θ [91].

The resulting response is given by [91]:

$$\Delta U = \frac{\beta \omega V_a^2 \sin \vartheta}{4V_{gt}|\sin kL|^2\sqrt{\omega^2 + \gamma^2}} \tag{3.33}$$

Here, ω is the frequency of the impinging THz radiation, k is the wave vector of a plasma wave:

$$k = (\Omega + i\Gamma)/s \tag{3.34}$$

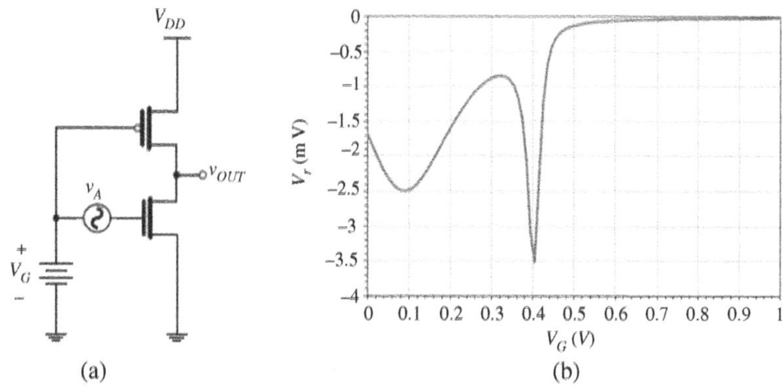

Figure 3.33 (a) A TFT complementary THz detector schematic diagram and (b) response with 1-segment TFT SPICE model. Parameters used in the simulation: $L_N = 0.8\,\mu m$, $L_P = 0.8\,\mu m$, $W_N = 250\,\mu m$, $W_P = 250\,\mu m$, $V_{tn} = 0.4\,V$, $V_{tp} = -0.4\,V$, $V_{DD} = 1\,V$, $C_{frs} = C_{frd} = C_{fdg} = 2.7\,fF$, THz amplitude: 10 mV. The THz source resistance of $50\,\Omega$. Here L_N, L_P, and W_N are the channel lengths and widths for n-channel and p-channel FETs, respectively. V_{tn}, V_{tp} are threshold voltages of n- and p-channel TFTs, V_{DD} is the power supply. Source: Shur et al. [79]/IOP Publishing/CC BY 4.0.

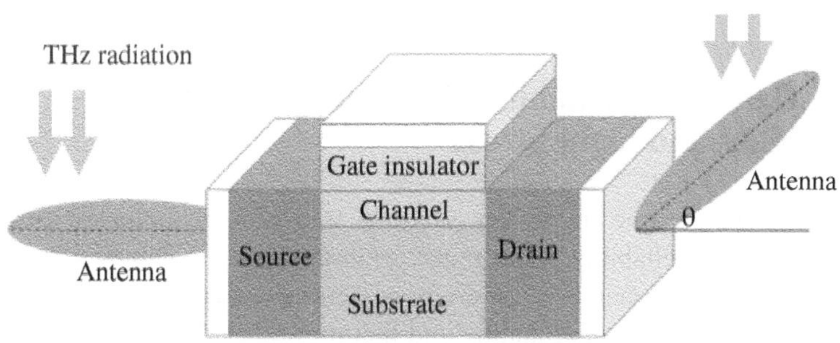

Figure 3.34 FET driven by signals from two antennas with the signals shifted by phase angle θ. Source: Adapted from Gorbenko et al. [91].

Here,

$$\Omega = \sqrt{\frac{\sqrt{\omega^4 + \omega^2\gamma^2}}{2} + \frac{\omega^2}{2}} \tag{3.35}$$

$$\Gamma = \sqrt{\frac{\sqrt{\omega^4 + \omega^2\gamma^2}}{2} - \frac{\omega^2}{2}} \tag{3.36}$$

s is the plasma velocity and γ is the collision frequency.

Figure 3.35 shows the response of a complementary inverter THz TFT detector as a function of frequency [16]. This dependence could be used to

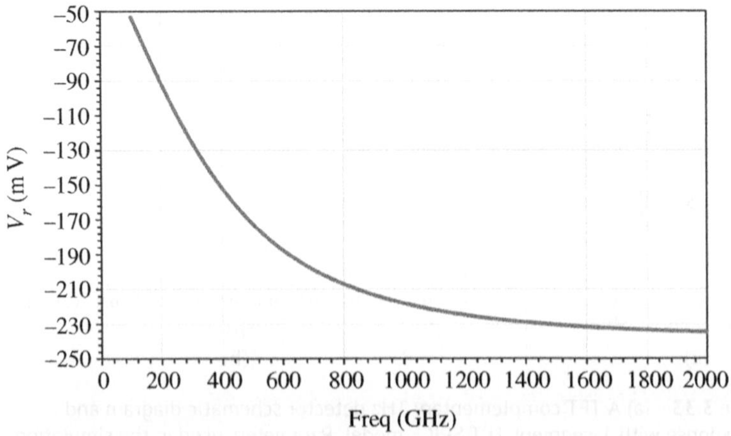

Figure 3.35 Response of a complementary inverter circuit for the THz TFT detector as a function of frequency. Source: Shur et al. [16]/IOP Publishing.

identify the frequency of impinging radiation operating oxide TFTs as spectrometers of THz radiation.

3.6 Oxide TFT Reliability, Durability, and Stability Under Bias and Light Stress

Reliability and lifetime are the key issues for transistor technologies, especially for TFTs. Reliability issues make it more difficult to deploy organic TFT technology commercially, for example. Generally speaking, IGZO TFTs have reached the level of reliability and lifetime sufficient for commercial applications. However, many problems remain. Oxide TFT's performance and lifetime are affected by issues related to spatial nonuniformities and electrically and illumination-induced instabilities.

The standard approach to estimating transistor lifetime is measuring lifetime at elevated temperatures, extracting activation energy, and extrapolating the temperature dependence of the high-temperature lifetimes to operating temperatures. However, such an analysis of the lifetime of wide-band semiconductor transistors (AlGaN/GaN HEMTs) showed that this approach could be inaccurate. Different degradation mechanisms might dominate at operating and higher temperatures, and activation energy is temperature-dependent, as shown in Figure 3.36 [97]. Therefore,

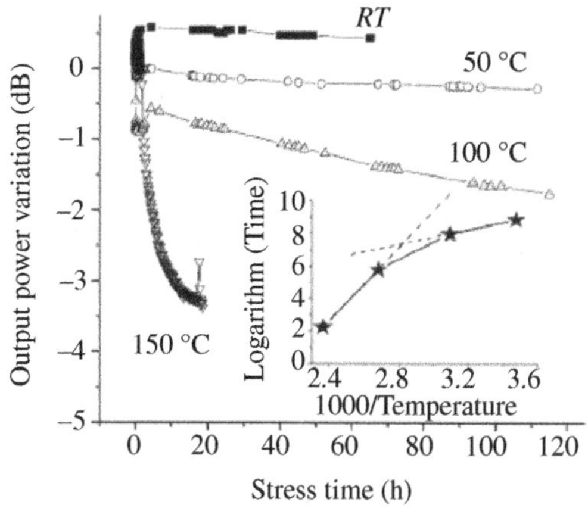

Figure 3.36 Power degradation during the stress test. Inset: Arrhenius plot describing power degradation. Dashed lines represent the range of activation energies. Source: Koudymov et al. [97]/John Wiley & Sons.

the activation energy temperature dependence for wide bandgap TFTs requires further investigation.

Other parameters characterizing reliability include the threshold voltage dependence on temperature and bias [98]. These dependencies should be accounted for in the compact modeling of oxide TFTs, just as it was done for amorphous silicon TFTs [99].

Just like in Si MOSFETs, threshold voltage shift under negative gate voltage is an important degradation mechanism in oxide TFTs. Depending on test conditions this voltage stress is called negative bias stress (NBS), negative bias illumination stress (NBIS), and negative bias temperature instability (NBTI). The mechanisms involved are quite similar. A negative bias induces trapped positive charges at the channel interface with a barrier layer and/or in the barrier layer. These trapped charges shift the threshold voltage. This process takes time, and the resulting threshold voltage shift increases with time. The problem is more severe in devices operating at smaller operating voltages. When the stress is removed, the threshold voltage shift gradually recovers. This recovery varies from device to device and depends on operating temperature making it difficult to account for induced threshold voltage shifts and the threshold voltage recovery in a circuit design. This irregularity is reflected in the term Instability and the letter I in the NBTI acronym.

The stress gate bias and temperature dependencies of the threshold voltage for NBS, NBIS, and NBTI degradation mechanisms and the recovery processes are similar [98]. The threshold voltage shift dependence on temperature is characterized by activation energy, E_a. It depends exponentially

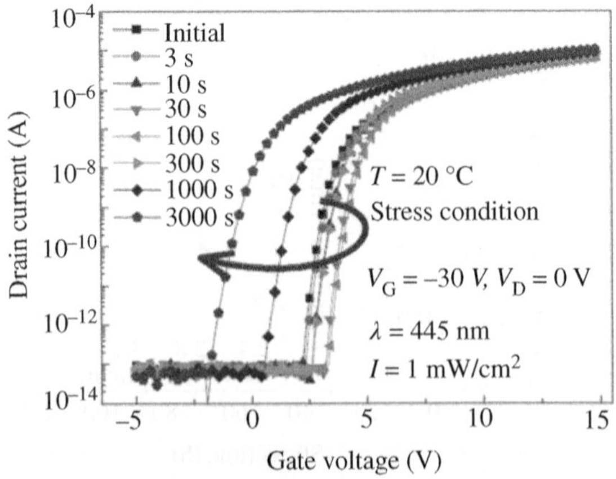

Figure 3.37 Transfer curves of the a-IGZO TFT vs. stress time under NBIS for $V_G = -30$ V. Source: Li et al. [105]/IEEE/CC BY 4.0.

on the gate bias V_G and stress time t, as shown below [100–104]:

$$\Delta V_{th} = \alpha t^n \exp\left(-\frac{E_a}{k_B T}\right) \exp(\beta V_G) \tag{3.37}$$

Stress conditions, such as gate-bias voltage and light illumination, also characterize the stability of oxide TFTs. It could be combined with temperature stress as well. The bias temperature stress of IGZO TFTs resulted in the threshold voltage shifts up to a few volts V, as shown in Figure 3.37 [105]. The threshold voltage shift is positive during the first 10 seconds of bias stress and then becomes negative and much larger at longer times.

The performance of a-IGZO TFTs is affected by light. Figure 3.38 shows (a) a schematic diagram of an a-IGZO TFT under light illumination, (b) transfer characteristics in the dark and after radiation with monochromatic light of varying wavelength (λ) for 500 s, and (c) transfer characteristics and μ_{FE} as a function of exposure time to deep UV light ($\lambda = 172$ nm, intensity of 400 mW/cm^2) [105]. Deep ultraviolet radiation ($\lambda = 175$ nm) induced a negative threshold-voltage shift and increased the subthreshold-voltage slope. These changes were linked to the creation and ionization of oxygen vacancy states (V_O), at the surface of the a-IGZO film. The transition from the neutral to the ionized V_O led to lattice relaxation. These results identified photo creation and ionization of V_O as the light instability mechanism in a-IGZO TFTs [105].

However, stable logic circuit operations have been achieved even for low-temperature IGZO TFTs [106]. Dielectric layers play an important role in parasitic effects in oxide TFT with some of the oxide layers resulting in voltage hysteresis. Using scandium zirconium oxide in fully printed oxide TFTs allowed achieving a negligible hysteresis [36].

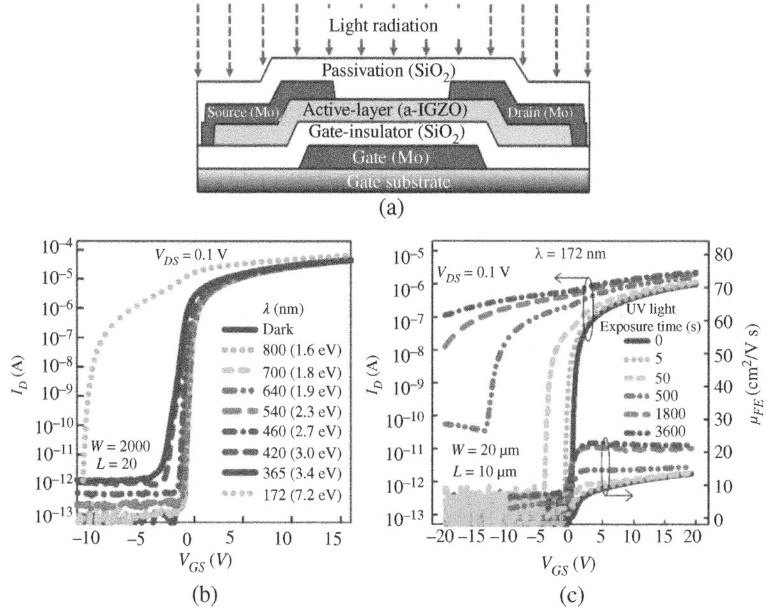

Figure 3.38 (a) Schematic of an a-IGZO TFT under light illumination, (b) transfer characteristics in dark and after radiation with lights of varying wavelengths for 500 s, (c) transfer characteristics and μ_{FE} as a function of exposure time to $\lambda = 172$ nm light at 400 mW/cm^2 intensity. Source: Adapted from Li et al. [105].

Back-channel graded nitrogen doping was shown to improve device performance and bias stress stability of IGZO TFTs [104]. Another technique to improve stability is using an inverted staggered etch stopper structure [107].

3.7 Summary

Oxide TFT technology could challenge a-Si and poly-Si TFT technology, organic flexible and wearable electronics technology, and even could further develop to challenge mainstream Si MOSFET Si CMOS technology for many applications. Oxide TFTs are even predicted to invade the sub-THz and THz frequency ranges. Field effect mobilities of 345 cm^2/V s have already been achieved but higher values are predicted allowing further shrinking of the performance gap between TFTs and Si CMOS. New fabrication methods such as high-temperature annealing enhance the electrical performance of the devices. The integration of oxide TFTs with flexible substrates and the development of roll-to-roll processing will support large-area flexible displays. More detailed discussions of the oxide TFT fabrication process and applications are included in Chapters 4 and 5, separately.

References

1 Lilienfeld, J. E. (1926). Method and apparatus for controlling electric currents. U. S. Patent No. 1,745,175, filed 8 October 1926 and issued 18 January 18 1930.

2 Lilienfeld, J. E. Wikipedia. Julius Edgar Lilienfeld (accessed 07 August 2022).

3 Shur, M.S., (1996). *Introduction to Electronic Devices*. New York: John Wiley and Sons.

4 Shockley, W. (1950). *Electrons and Holes in Semiconductors with Applications to Transistor Electronics*. New York: Van Nostrand.

5 Luryi, S. (1988). Quantum capacitance devices. *Applied Physics Letters* 52 (6): 501–503.

6 Cerebras Systems. www.cerebras.net/company/news/ (accessed 26 November 2023).

7 Wallmark, J. T. (1957). Field-effect transistor. U.S. Patent 2,900,531 (1959), filed 28 February 1957 and issued 8 August 1959.

8 Nomura, K. (2021). Recent progress of oxide-TFT-based inverter technology. *Journal of Information Display* 22 (4): 211–229.

9 Jeon, Y., Lee, D., and Yoo, H. (2022). Recent advances in metal-oxide thin-film transistors. Flexible/stretchable devices, integrated circuits, biosensors, and neuromorphic applications. *Coatings* 12 (2): 204.

10 Lyu, R.J., Shie, B.S., Lin., H. C. et al. (2017). Downscaling metal-oxide thin film transistors to sub-50 nm in an exquisite film-profile engineering approach. *IEEE Transactions on Electron Devices* 64 (3): 1069.

11 Shih, C.W., Chin, A., Lu, C.F. et al. (2018). Remarkably high hole mobility metal-oxide thin film transistors. *Science Reports* 8: 1.

12 Rim, Y.S., Chen, H., Zhu, B. et al. (2017). Interface engineering of metal oxide semiconductors for biosensing applications. *Advanced Material Interfaces* 4 (10): 1700020.

13 Dmitriev, A.P. and Shur, M.S. (2006). Ballistic admittance: periodic variation with frequency. *Applied Physics Letters* 89 (14): 142102.

14 Kastalsky, A.A. and Shur, M.S. (1981). Conductance of small semiconductor devices. *Solid State Communications* 39 (6): 715–718.

15 Shur, M.S. (2002). Low ballistic mobility in submicron high electron mobility transistors. *IEEE Electron Device Letters* 23 (9): 511–513.

16 Shur, M., Liu, X., and Ytterdal, T. (2022). Improved thin film transistor model predicts TFT operation in the THz range. *2022 ECS Transactions* 109 (3): 3–12.

17 Yang, X.Z., Liu, X.-D., and Li, Z.-Q. (2020). Electron effective mass and electronic structure in nonstoichiometric amorphous indium gallium zinc oxide films. *Journal of Alloys and Compounds* 813 (1): 152183.

18 Mukherjee, A., Ottapilakkal, V., Sagar, S. et al. (2021). Ultralow-voltage field-effect transistors using nanometer-thick transparent amorphous indium–gallium–zinc oxide films. *ACS Applied Nano Materials* 4: 8050–8058.

19 Naqi, M., Cho, Y., and Kim, S. (2023). High-speed current switching of inverted-staggered bottom-gate a-IGZO-based thin-film transistors with highly stable logic circuit operations. *ACS Applied Electronic Materials* 5 (6): 3378–3383.

20 Chiu, C.J., Chang, S.P., and Chang, S.J. (2010). High-performance a-IGZO thin-film transistor using Ta_2O_5 gate dielectric. *IEEE Electron Device Letters* 31 (11): 1245–1247.

21 Sheng, J., Hong, T., Lee, H.M. et al. (2019). Amorphous IGZO TFT with high mobility of approximately 70 cm^2/Vs via vertical dimension control using PEALD. *ACS Applied Material Interfaces* 11 (43): 40300–40309.

22 Samanta, S., Chand, U., Xu, S. et al. (2020). Low subthreshold swing and high mobility amorphous indium–gallium–zinc-oxide thin-film transistor with thin HfO_2 gate dielectric and excellent uniformity. *IEEE Electron Device Letters* 41 (6): 856–859.

23 Swallow, J.E.N. (2021). Indium gallium oxide alloys: electronic structure, optical gap, surface space charge, and chemical trends within common-cation semiconductors. *ACS Applied Material Interfaces* 13 (2): 2807–2819.

24 Bong, H., Lee, W.H., Lee, D.Y. et al. (2010). High-mobility low-temperature ZnO transistors with low-voltage operation. *Applied Physics Letters* 96 (19): 192115.

25 Chen, X., Zhang, G.Z., Wan, J.X. et al. (2019). Transparent and flexible thin-film transistors with high performance prepared at ultralow temperatures by atomic layer deposition. *Advanced Electronic Materials* 5 (2): 1800583.

26 Wang, M., Zhan, D., Wang, X. et al. (2021). Performance optimization of atomic layer deposited ZnO thin-film transistors by vacuum annealing. *IEEE Electron Device Letters.* 42 (5): 716–719.

27 Li, S., Chen, X., Liu, L. et al. (2022). Micron channel length ZnO thin film transistors using bilayer electrodes. *Journal of Colloid and Interface Science* 622: 769–779.

28 Zhao, W., Zhang, N., Yao, C. et al. (2023). Enhanced stability performance of transparent ozone ALD ZnO thin-film transistors with $SiAlO_x$ dielectric. *IEEE Trans. Electron Devices* 70 (2): 556–562.

29 Si, M., Lin, Z., Charnas, A. et al. (2020). Scaled atomic-layer-deposited indium oxide nanometer transistors with maximum drain current exceeding 2 A/mm at drain voltage of 0.7 V. *IEEE Electron Device Letters* 42 (2): 184–187.

30 Si, M., Hu, Y., Lin, Z. et al. (2021). Why In_2O_3 can make 0.7 nm atomic layer thin transistors. *Nano Letters* 21 (1): 500–506.

31 Liang, L.Y., Liu, Z.M., Cao, H.T. et al. (2010). Improvement of phase stability and accurate determination of optical constants of SnO thin films by using Al_2O_3 capping layer. *ACS Applied Materials and Interfaces* 2 (6): 1565–1568.

32 Yao, Z.Q., Liu, S.L., Zhang, L. et al. (2012). Room temperature fabrication of p-channel Cu_2O thin-film transistors on flexible polyethylene terephthalate substrates. *Applied Physics Letters* 101 (4): 042114.

33 Sun, H., Chen, S.C., Wen, C.K. et al. (2017). P-type cuprous oxide thin films with high conductivity deposited by high power impulse magnetron sputtering. *Ceramics International* 43 (8): 6214–6220.

34 Ouyang, Z., Wang, W., Dai, M. et al. (2022). Research progress of p-type oxide thin-film transistors. *Materials (Basel)* 15 (14): 4781.

35 Liu, A., Zhu, H., Park, W.T. et al. (2020). High-performance p-channel transistors with transparent Zn doped-CuI. *Nature Communications* 11: 4309.

36 Li, Y., Lan, L., Hu, S. et al. (2019). Fully printed top-gate metal–oxide thin-film transistors based on scandium-zirconium-oxide dielectric. *IEEE Transactions on Electron Devices* 66 (1): 445–450.

37 Liang, L.Y., Cao, H.T., Chen, X.B. et al. (2012). Ambipolar inverters using SnO thin-film transistors with balanced electron and hole mobilities. *Applied Physics Letters* 100: 263502.

38 Shang, Z.-W., Hsu, H.-H., Zheng, Z.-W. et al. (2019). Progress and challenges in p-type oxide-based thin film transistors. *Nanotechnology Reviews* 8 (1): 422–443.

39 Si, M., Charnas, A., Lin, Z. et al. (2021). Enhancement-mode atomic layer-deposited In_2O_3 transistors with maximum drain current of 2.2 A/mm at drain voltage of 0.7 V by low-temperature annealing and stability in hydrogen environment. *IEEE Trans. Electron Devices* 68 (3): 1075–1080.

40 Chang, S.-W., Lu, T.H., Yang, C.Y. et al. (2022). First demonstration of heterogeneous IGZO/Si CFET monolithic 3-D integration with dual work function gate for ultralow-power SRAM and RF applications. *IEEE Transactions on Electron Devices* 69 (4): 2101–2107.

41 Inoue, H., Matsuzaki, T., Nagatsuka, S. et al. (2012). Nonvolatile memory with extremely low-leakage indium-gallium-zinc-oxide thin-film transistor. *IEEE Journal of Solid-State Circuits* 47 (9): 2258–2265.

42 Rockelé, M., Pham, D.-V., Hoppe, A. et al. (2011). Low-temperature and scalable complementary thin-film technology based on solution-processed metal oxide n-TFTs and pentacene p-TFTs. *Organic Electronics* 12 (11): 1909–1913.

43 Xu, Q., Zhao, J., Pecunia, V. et al. (2017). Selective conversion from p-type to n-type of printed bottom-gate carbon nanotube thin-film transistors and application in complementary metal–oxide–semiconductor inverters. *ACS Applied Materials and Interfaces* 9 (14): 12750–12758.

44 Faber, H., Das, S., Lin, Y.-H. et al. (2017). Heterojunction oxide thin-film transistors with unprecedented electron mobility grown from solution. *Science Advances* 3: e1602640.

45 Noh, J.H., Ryu, S.Y., Jo, S.J. et al. (2010). Indium oxide thin-film transistors fabricated by RF sputtering at room temperature. *IEEE Electron Device Letters* 31 (6): 567–569.

46 Hirao, T., Furuta, M., Furuta, H. et al. (2007). Novel top-gate zinc oxide thin-film transistors (ZnO TFTs) for AMLCDs. *Journal of Society for Information Display* 15 (1): 17–22.

47 Chiu, I.-C. and Cheng, I.-C. (2013). Gate-bias stress stability of p-type SnO thin-film transistors fabricated by RF-sputtering. *IEEE Electron Device Letters* 35 (1): 90–92.

48 Cho, M.H., Seol, H., Song, A. et al. (2019). Comparative study on performance of IGZO transistors with sputtered and atomic layer deposited channel layer. *IEEE Transactions on Electron Devices* 66 (4): 1783–1788.

49 Saha, J.K., Bukke, R.N., Mude, N.N. et al. (2020). Significant improvement of spray pyrolyzed ZnO thin film by precursor optimization for high mobility thin film transistors. *Science Reports* 10: 8999.

50 Fortunato, E., Pimentel, A., Pereira, L. et al. (2004). High field-effect mobility zinc oxide thin film transistors produced at room temperature. *Journal of Non-Crystalline Solids* 338–340 (1): 806–809.

51 Cho, M.H., Choi, C.H., Seul, H.J. et al. (2021). Achieving a low-voltage, high-mobility IGZO transistor through an ALD-derived bilayer channel and a hafnia-based gate dielectric stack. *ACS Applied Materials & Interfaces* 13 (14): 16628–16640.

52 Magari, Y., Kataoka, T., Yeh, W. et al. (2022). High-mobility hydrogenated polycrystalline In_2O_3 (In_2O_3:H) thin-film transistors. *Nature Communications* 13: 1078.

53 Su, L.-Y., Lin, H.-Y., Lin, H.-K. et al. (2011). Characterizations of amorphous IGZO thin-film transistors with low subthreshold swing. *IEEE Electron Device Letters* 32 (9): 1245–1247.

54 Lin, Z., Wang, Z., Zhao, J. et al. (2023). A low-leakage zinc oxide transistor by atomic layer deposition. *IEEE Electron Device Letters* 44 (3): 536–539.

55 Charnas, A., Lin, Z., Zhang, Z. et al. (2021). Atomically thin In_2O_3 field-effect transistors with 10^{17} current on/off ratio. *Applied Physics Letters* 119 (26): 263503.

56 Kim, H.-M., Choi, S.-H., Lee, H.U. et al. (2023). The significance of an in situ ALD Al_2O_3 stacked structure for p-type SnO TFT performance and monolithic all-ALD-channel CMOS inverter applications. *Advanced Electronic Materials* 9 (4): 2201202.

57 Nomura, K., Kamiya, T., and Hosono, H. (2011). Ambipolar oxide thin-film transistor. *Advanced Materials* 23 (30): 3431.

58 Huang, C.-H., Zhang, Y., and Nomura, K. (2022). Reconfigurable artificial synapses with excitatory and inhibitory response enabled by an ambipolar oxide thin-film transistor. *ACS Applied Materials and Interfaces* 14 (19): 22252–22262.

59 Yarali, E., Faber, H., Yengel, E. et al. (2020). Low-voltage heterojunction metal oxide transistors via rapid photonic processing. *Advanced Electronic Materials* 6 (6): 2000028.

60 Jeong, J.W., Hwang, H.S., Choi, D. et al. (2020). Hybrid polymer/metal oxide thin films for high performance, flexible transistors. *Micromachines* 11 (3): 264.

61 Choi, S.-H. (2021). High-performance oxide tfts with co-sputtered indium tin oxide and indium-gallium-zinc oxide at source and drain contacts. *IEEE Electron Device Letters* 42 (2): 168–171.

62 Chen, T.-Y., Yang, T.-H., Wu, N.-T. et al. (2017). Transient analysis of streptavidin-biotin complex detection using an IGZO thin film transistor-based biosensor integrated with a microfluidic channel. *Sensors and Actuators B: Chemical* 244: 642–648.

63 Wang, Y.-W., Chen, T.-Y., Yang, T.-H. et al. (2016). Thin-film transistor-based biosensors for determining stoichiometry of biochemical reactions. *PLOS One* 11 (12): e0169094.

64 Wang, D., Liu, Z., Li, J. et al. (2022). Thin-film transistor arrays for biological sensing systems. *Flexible and Printed Electronics* 7 (2): 023004.

65 Shur, M.S. (1984). Folded gate-A novel logic gate structure. *IEEE Electron Device Letters* 5 (11): 454–455.

66 Nguyen, A.H.T., Nguyen, M.C., Cho, S. et al. (2020). Double-gate thin film transistor with suspended-gate applicable to tactile force sensor. *Nano Convergence* 7: 31.

67 Li, Q., Dong, J., Han, D. et al. (2022). Back-end-of-line compatible InSnO/ZnO heterojunction thin-film transistors with high mobility and excellent stability. *IEEE Electron Device Letters* 43 (8): 1251–1254.

68 Mukhopadhyaya, K. and Srividya, P. (2022). Trends in performance characteristics and modelling of oxide based TFT. *Materials Today: Proceedings* 55 (2): 414–418.

69 Deng, W., Zhang, W., Peng, Y. et al. (2021). Compact modeling of metal-oxide TFTs based on artificial neural network and improved particle swarm optimization. *Journal of Computational Electronics* 20 (4): 1043–1049.

70 Keysight Technologies. www.keysight.com/us/en/home.html (accessed 10 December 2023).

71 PSPICE. www.pspice.com (accessed 10 December 2023).

72 Martins, J., Barquinha, P., and Goes, J. (2016). TCAD simulation of amorphous -gallium-zinc oxide thin-film transistors. In: *Technological Innovation for Cyber-Physical Systems, IFIP Advances in Information and Communication Technology*, vol. 470 (ed. L.M. Camarinha-Matos, A.J. Falcão, N. Vafaei, and S. Najdi). Springer.

73 Abe, K., Ota, K., and Kuwagaki, T. (2020). Device modeling of oxide semiconductor TFTs. In: *2020 27th International Workshop on Active-Matrix Flat Panel Displays and Devices (AM-FPD)*, 137–140. Japan: Kyoto.

74 Liu, X.T., Ytterdal, T., and Shur, M. (2020). TeraFET amplifier detector. In: *2020 International Topical Meeting on Microwave Photonics (MWP)*, 216–219. Japan: Matsue.

75 Liu, X.T., Ytterdal, T., and Shur, M. (2021). Line of sight THz detector using TeraFET spectrometers. In: *2021 46th International Conference on Infrared, Millimeter and Terahertz Waves (IRMMW-THz), 2021*, 1–2.

76 Liu, X.T., Ytterdal, T., and Shur, M. (2022). Multi-segment TFT compact model for THz applications. *Nanomaterials* 12 (5): 765.

77 Liu, X., Dovidenko, K., Park, J. et al. (2018). Compact terahertz SPICE model: effects of Drude inductance and leakage. *IEEE Transactions on Electron Devices* 65 (12): 5350.

78 Gundlach, D.J., Zhou, L., Nichols, J.A. et al. (2006). An experimental study of contact effects in organic thin film transistors. *Journal of Applies Physics* 100: 024509.

79 Shur, M.S., Liu, X., and Ytterdal, T. (2023). Thin film transistor response in the THz range. *ECS Journal of Solid State Science and Technology* 12: 035008.

80 Shur, M. (2010). Plasma wave terahertz electronics. *Electronics Letters* 46 (26): 18–21.

81 Shur, M.S. (1981). Ballistic transport in semiconductor with collisions. *IEEE Transactions on Electron Devices* 28 (10): 1120.

82 Ghediya, P., Yang, H., Fujimoto, T. et al. (2023). Improved electron transport properties of Zn-Rich In–Ga–Zn–O thin-film transistors. *Journal of Physical Chemistry C* 127 (5): 2622–2627.

83 Dyakonov, M.I. and Shur, M.S. (1996). Plasma wave electronics: novel terahertz devices using two-dimensional electron fluid. *IEEE Transactions on Electron Devices* 43 (10): 1640.

84 Shur, M., Ryzhii, V., Gaska, R. (2009). Method of radiation generation and manipulation. U.S. Patent 7,619,263), filed 17 November 2009 an issued 11 March 2010.

85 Matsubayashi, D., Asami, Y., Okazaki, Y. et al. (2015). 20-nm-node trench-gate-self-aligned crystalline In-Ga-Zn-Oxide FET with high frequency and low off-state current. In: *2015 IEEE International Electron Devices Meeting (IEDM)*, vol. 2015, 6.5.1–6.5.4. Washington, DC, USA.

86 Aizin, G.R., Mikalopas, J., and Shur, M. (2016). Current driven plasmonic boom instability in gated periodic ballistic nanostructures. *Physical Review B* 93 (19): 195315.

87 Liu, X., Ytterdal, T., Kachorovskii, V.Y. et al. (2019). Compact terahertz SPICE/ADS model. *IEEE Transactions on Electron Devices* 66 (6): 2496.

88 Liu, X. T., Ytterdal, T., Shur, M. (2021). Design and optimization of TeraFET spectrometer. *8th International IEEE Conference on Microwaves, Communications, Antennas, Biomedical Engineering and Electronic Systems (COMCAS)*. Tel Aviv, Israel, 1–3 November 2021; pp. 477–481.

89 Gutin, A., Ytterdal, T., Muraviev, A. et al. (2015). Modeling effect of parasitics in plasmonic FETs. *Solid State Electronics* 104: 75–78.

90 Liu, X.T., Ytterdal, T., and Shur, M. (2022). Optimization of Si CMOS TeraFETs for 300 GHz band operation. In: *47th International Conference on Infrared, Millimeter and Terahertz Waves (IRMMW-THz)*, 1–2.

91 Gorbenko, I.V., Kachorovskii, V.Y., and Shur, M. (2019). Terahertz plasmonic detector controlled by phase asymmetry. *Optics Express* 27 (4): 4004–4013.

92 Liu, X., Ytterdal, T., and Shur, M. (2020). Plasmonic FET terahertz spectrometer. *IEEE Access* 99 (3): 1–6.

93 Shur, M. (2023). Phase-sensitive silicon CMOS TeraFETs. In: *48th International Conference on Infrared, Millimeter, and Terahertz Waves (IRMMW-THz)* (ed. Q.C. Montreal), 1–2. Canada.

94 Shur, M. (2023). Heterodyne and phase sensitive plasmonic terahertz detectors and spectrometers. *In Encyclopedia of Materials: Electronics (Volume III)* 0128197285, 9780128197288: 97–103

95 Liu, X. T., Ytterdal, T., Shur. M. (2021). Traveling wave TeraFET spectrometer. *46th International Conference on Infrared, Millimeter and Terahertz Waves (IRMMW-THz)*, pp. 1–2. DOI: https://doi.org/10.1109/IRMMW-THz50926.2021.9567354.

96 Liu, X., Ytterdal, T., and Shur, M. (2021). Frequency to digital conversion using Si TeraFETs. *Optical Engineering* 60 (8): 082017-1-8.

97 Koudymov, A., Shur, M.S., Simin, G. et al. (2007). Current collapse and reliability of III-N heterostructure field effect transistors. *Physica Status Solidi (RRL)-Rapid Research Letters* 1 (3): 116–118.

98 Martins, J., Bahubalindruni, P., Rovisco, A. et al. (2017). Bias stress and temperature impact on InGaZnO TFTs and circuits. *Materials (Basel)* 10 (6): 680.

99 Iniguez, B., Wang, L., Xu, Z. et al. (1999). Short channel AIM-Spice models for amorphous and polysilicon thin film transistors. In: *Future Trends in Microelectronics. The Road Ahead* (ed. S. Luryi, J. Xu, and A. Zaslavsky), 213–220. Wiley and Sons.

100 Krishnan, A.T., Reddy, V., and Krishnan, S. (2001). Impact of charging damage on negative bias temperature instability. In: *International Electron Devices Meeting. Technical Digest (Cat. No.01CH37224)*, 39.3.1–39.3.4. Washington, DC, USA: https://doi.org/10.1109/IEDM.2001.979650.

101 Chen, C.-Y., Lee, J.-W., Wang, S.-D. et al. (2006). Negative bias temperature instability in low temperature polycrystalline silicon thin-film transistors. *IEEE Transactions on Electron Devices* 53 (12): 2993–3000.

102 Bak, J., Kang, Y., Yang, S. et al. (2015). Origin of degradation phenomenon under drain bias stress for oxide thin film transistors using IGZO and IGO channel layers. *Science Reports* 5: 7884.

103 Mativenga, M., Haque, F., Billah, M.M. et al. (2021). Origin of light instability in amorphous IGZO thin-film transistors and its suppression. *Science Reports* 11: 14618.

104 Zhu, Z., Cao, W., Huang, X. et al. (2022). Analysis of nitrogen-doping effect on sub-gap density of states in a-IGZO TFTs by TCAD simulation. *Micromachines (Basel)* 13 (4): 617.

105 Li, S., Wang, M., Zhang, D. et al. (2019). A unified degradation model of a-InGaZnO TFTs under negative gate bias with or without an illumination. *Journal of the Electron Devices Society* 7: 1063–1071.

106 Mativenga, M., Choi, J.W., Hur, J.H. et al. (2011). Highly stable amorphous indium–gallium–zinc-oxide thin-film transistor using an etch-stopper and a via-hole structure. *Journal of Information Display* 12 (1): 47–50.

107 Kim, C.H., Rim, Y.S., and Kim, H.J. (2014). The effect of a zinc-tin-oxide layer used as an etch-stopper layer on the bias stress stability of solution-processed indium-gallium-zinc-oxide thin-film transistors. *Journal of Physics D: Applied Physics* 47 (38): 385104.

4

Oxide Thin Film Transistor Fabrication Processes

4.1 Introduction

The hydrogenated amorphous silicon (a-Si:H), low-temperature poly-Si (LTPS), and oxide TFT backplanes are being used for the manufacturing of display products. A-Si:H TFT backplane could be deposited on G10.5 and G11 glass substrates, but G6.5 glass is the maximum size for an LTPS TFT backplane. This makes LTPS expensive for display products. In principle, the largest glass size for fabricating the oxide TFT backplanes can be the same as that used in a-Si:H TFT backplanes, i.e., G10.5 or G11. Therefore, backplanes made of oxide TFTs are cheaper than those made of LTPS TFTs. Currently, oxide TFT backplanes for OLEDs are produced on G8 glass substrates for TVs. An important difference between oxide TFT and LTPS TFT manufacturing processes is the low-temperature, large-area capability. Active-matrix OLED (AMOLED) displays using oxide TFT backplanes are being diversified from TVs to tablet and laptop monitors.

Reactive sputtering is a common method for depositing oxide semiconductor (OS) thin films. The film's electrical properties, such as Hall mobility and carrier concentration, are very sensitive to the deposition condition, such as the O_2 and H_2 partial pressures or the substrate temperature. The OS surface can be treated with the N_2O plasma to achieve a high resistivity. Doping is important for various oxide devices. For example, IGZO can be treated with Ar, He, H_2, SiH_4, NF_3, or CF_4 plasma to obtain the n^+ contact layer.

The solution process has been used to prepare OS materials at a low cost without requiring the vacuum condition. Spin coating is widely used, but the film's uniformity is not satisfactory. On the other hand, spray pyrolysis (SP) can provide good film uniformity over a large area. Polycrystalline/nanocrystalline oxide semiconductor films, such as IGZO and $In_{0.8}Ga_{0.2}O$,

Oxide Thin Film Transistors, First Edition. Yue Kuo, Hideo Hosono, Michael S. Shur, and Jin Jang.

can be prepared by SP at a relatively high temperature, such as 420 °C. Inkjet printing can be used to selectively deposit the oxide semiconductor film on the TFT channel region. However, the performance and stability of the TFT are not as good as those of the sputtered deposited TFT.

Oxide TFTs can have the inverted staggered, e.g., etch stopper (ES) or back channel etch (BCE), or coplanar structure. The ES TFT can show good performance because the ES layer protects the active semiconductor layer during the TFT process. The structure of BCE TFT is the same as that of a-Si:H TFT which is currently used for manufacturing AMLCD. However, both structures have large parasitic capacitance, which is a disadvantage for a large-area display. The coplanar oxide TFT has negligible parasitic capacitance and is currently used for large-size OLED displays, such as TVs.

The plasma enhanced chemical vapor deposition (PECVD) SiO_2 is the most popular gate insulator (GI) in mass-produced oxide TFTs. On the other hand, Al_2O_3 prepared by atomic layer deposition (ALD) or PE-ALD has been widely explored for gate dielectric application. When SiO_2 is used as the gate dielectric, it is usually used as an interlayer between the gate and source/drain (S/D). PECVD SiN_x is an excellent gas barrier deposited on top of the oxide TFT array. Separately, the multilayer (SiO_2/SiN_x) stack is used as the gas barrier for polyimide (PI) to block the penetration of water vapor and O_2 from the air.

The performance of an oxide TFT is very sensitive to the process condition and device structure. This chapter summarizes the influence of process conditions on the properties of oxide TFTs. Since the O vacancy (V_o), which can be a donor, can be easily generated during the fabrication process, methods of reducing its generation, such as the N_2O plasma treatment, are discussed.

4.2 Deposition Methods for Individual Layers in Oxide TFTs

4.2.1 Process Influences on Individual Layer Material Properties

Sputtering is a common deposition method for OS thin films. During sputtering the atoms are ejected from a solid target (oxide semiconductor) due to bombardment by energetic Ar ions. OS is usually deposited in a vacuum chamber by flowing Ar and O_2 to incorporate O in the OS [1]. The O_2 partial pressure is an important deposition parameter because the amount of O in OS increases with O_2 partial pressure [2, 3]. Therefore, the resistivity of OS increases with O_2 partial pressure. Note that the performance of OS TFT strongly depends on its resistivity [4].

Deposition temperature is also an important parameter to determine the OS properties such as density of states (DOS) in the gap, hydrogen (H) content, and mass density (ρ_m). The amorphous (a-) OS for TFTs is usually deposited at room temperature (RT) to 200 °C in a sputter chamber [5]. But the OS thin film deposited at higher substrate temperature, for example over 300 °C, can have more O vacancies. As a result, the threshold voltage (V_{th}) of OS TFT can be negative, having a depletion mode transfer curve of the TFT [6].

A disadvantage of OS TFT is the controllability of V_{th} because there is no good method for precise control of its conductivity. The precise control of V_{th} is important to design the CMOS circuits. On the other hand, the V_{th} can be precisely controlled by channel doping in the active layer of LTPS TFTs [7].

H is incorporated in OS and this depends on the deposition conditions, such as base pressure, substrate temperature, and gas pressure during sputtering. H_2 can be incorporated with Ar into the deposition chamber during sputtering [8]. H_2 partial pressure in Ar controls the resistivity of a-IGZO. Note that suitable H_2 incorporation in Ar during the sputtering decreases the trap in IGZO. It is known that the H incorporation in IGZO increases carrier density due to its shallow donor behavior [9, 10]. The typical H content is ~10^{20} cm^{-3} in IGZO film deposited by reactive sputtering. It is important to note that, there are favorable and unfavorable effects of H on TFT performance [11]. The incorporation of H into the IGZO film reduces the metal–metal (M—M) related defect states near the conduction band minimum by forming metal–hydrogen (M—H) bonds [12]. In addition, the formation of M—H bonds is facilitated by the stabilization of H at the V_o site [13].

The M—H bonds, such as In-H, Ga–H, and Zn–H can be formed during hydrogenated amorphous silicon (a-Si:H) deposition on top of IGZO [14]. It is demonstrated that the highly stable, self-aligned coplanar a-IGZO TFTs can be achieved by the deposition of a-Si:H on the source/drain contact region of the a-IGZO at a high substrate temperature of 360 °C. This process helps to obtain n^+ region for ohmic contact. The sheet resistance of the n^+ a-IGZO S/D contact region is less than 1 kΩ/square using a 20–30 nm a-Si:H layer. The short channel length, coplanar a-IGZO TFT with less than 1 μm exhibits excellent performance, showing the subthreshold swing (SS) of 0.15 V/dec, V_{th} of 0.1 V, and field effect mobility (μ_{FE}) of over 10 cm^2/V s [14]. The TFTs also demonstrate stable performance under positive bias temperature stress (PBTS) at 60 °C for a constant bias stress of 20 V. The low resistivity and highly stable ohmic contact are obtained due to the presence of stable M—H bonds, which results from the deposition of a-Si:H on a-IGZO at 360 °C [14].

Plasma treatments with N_2O, Ar, H_2, NF_3, etc. are popular techniques to control the resistivity of OSs. Among them, N_2O plasma treatment is widely implemented in oxide TFT fabrication [15–20]. However, the process conditions, such as discharging power, deposition temperature, and treatment time, should be optimized for oxide TFTs. A key advantage of N_2O plasma treatment is that it can control the Fermi level by increasing the resistivity of IGZO. The V_{th} shifts towards positive by increasing N_2O plasma exposure time due to the increase of resistivity [21]. However, plasma-induced damage can exist on OS if the plasma power is high or the plasma exposure time is too long. This indicates that the optimum process condition for the treatment should be developed to obtain a good-quality OS. It is also reported that the interface state density (D_{it}), V_o, and OH concentrations are decreased by N_2O plasma treatment [22].

Annealing of OS thin-film under N_2O environment can reduce the traps in OS and its interface with GI, resulting in the improvement of TFT performance. Note that the V_{th} shifts to the negative direction by N_2O annealing due to the passivation of traps [23, 24]. The N_2O annealing can improve the IGZO TFT performance because of the passivation of defects with O radicals. It can be concluded that for OS TFTs, the N_2O annealing can shift the V_{th} towards the negative gate voltage direction but the N_2O plasma treatment can shift the V_{th} toward the positive gate voltage direction [25–27]. N_2O plasma treatment on the IGZO layer incorporates O into IGZO and thus the concentration of V_o could not be high even after deposition of the top SiO_2 GI and/or the passivation layer. This can suppress the change into depletion mode TFT during the GI and passivation processes.

Among the other plasma treatments, Ar plasma usually generates V_o on OS. Therefore, it can be used for obtaining n^+ OS for coplanar TFTs [28]. NF_3 plasma treatment can also be used to make n^+ OS because F incorporation leads to a decrease of the resistivity [29, 30]. In addition, UV exposure decreases the resistivity because of the generation of V_o in OS [31, 32]. Helium (He) plasma treatment can be also used to obtain ohmic contact in oxide TFTs because of the reduction of resistivity [33, 34]. The H_2 plasma exposure can reduce the resistivity of OS, which is suitable for ohmic contact with S/D metal electrodes [35, 36]. The deposition of SiN_x on OS can decrease the resistivity due to the H diffusion into the OS film [37]. The reaction between S/D metals, such as Al [38] and Mo [39], and OS can reduce the resistivity so that can be used for ohmic contact in the oxide TFT.

The chemical composition in OS thin film plays a major role in its device performance. X-ray Photoelectron Spectroscopy (XPS) is a typical method to analyze the metal-oxide (M—O) bond, V_o, and OH concentrations [40]. The high M-O concentration, low V_o, and OH concentrations are preferable for high-density OS films. Usually, V_o acts as a donor and OH leads to the generation of a trap state [41].

4.2.2 Gate Insulators and Passivation Layers for Oxide TFTs

The PECVD SiO_2 is widely used for the GI and passivation layer of oxide TFTs [42–45]. Note that the quality of PECVD SiO_2 depends on the substrate temperature, plasma discharging power, N_2O/SiH_4 ratio during SiO_2 deposition, and mass density [46, 47]. The SiO_2 deposited from N_2O and SiH_4 with He dilution can reduce the leakage current density [48]. SiO_2 deposited at a low temperature, e.g., <200 °C, has a significant amount of H, which leads to a high leakage current density [48]. However, the high temperature, e.g., >350 °C, deposited SiO_2 has a negligible amount of H, which can be widely used in LTPS and OS TFTs [48].

SiN_x deposited by PECVD at 350 °C has quite a large amount of H, e.g., >20 at.% [49]. This indicates that H in SiN_x can be diffused into OS during the TFT fabrication and/or post-fabrication annealing process. Therefore, high-temperature deposited SiN_x is used as the bottom GI for the coplanar dual gate oxide TFT. SiN_x is not used as the interlayer on OS due to its high H content [50]. Even when SiN_x is deposited at a high temperature (350–400 °C), a significant amount of highly stable H bonds exists in the film [51, 52]. SiN_x can be used as a gas barrier layer for coplanar oxide TFT. As the top passivation, it blocks the diffusion of water vapor from the air [53]. Since the water vapor transmission rate (WVTR) of SiN_x is much lower than that of SiO_2, it has been used as the passivation layer on OLED for the smartphone OLED displays.

The SiO_2/SiN_x stack layer is widely used as the GI in coplanar LTPS, bottom gate BCE, and ES oxide TFTs [54–56]. The interface between OS and SiO_2 has a low D_{it}, so that the *SS* is quite low, typically less than 0.2 V/dec. The PE-ALD Al_2O_3 can be used for GI and passivation layer, but the deposition rate of the ALD process is much lower than that of PECVD SiO_2 [57]. There are many reports on increasing the throughput of PE-ALD Al_2O_3 [57–60]. However, the ALD process cannot support the G6 and G8 glass substrates.

4.3 Fabrication Processes of Oxide TFTs with Various Structures

4.3.1 Etch-stop Oxide TFTs

The ES structure was utilized in the early stages of oxide TFT research for making display prototypes. It was particularly used for TFT backplanes for AMLCD and AMOLED displays [61–63]. The ES material is usually SiO_2 that is deposited on top of the OS before the deposition of source/drain metal electrodes, as shown in Figure 4.1. The fabrication process flow is: (1) bottom gate formation, (2) GI, OS, and ES layers deposition, (3) after

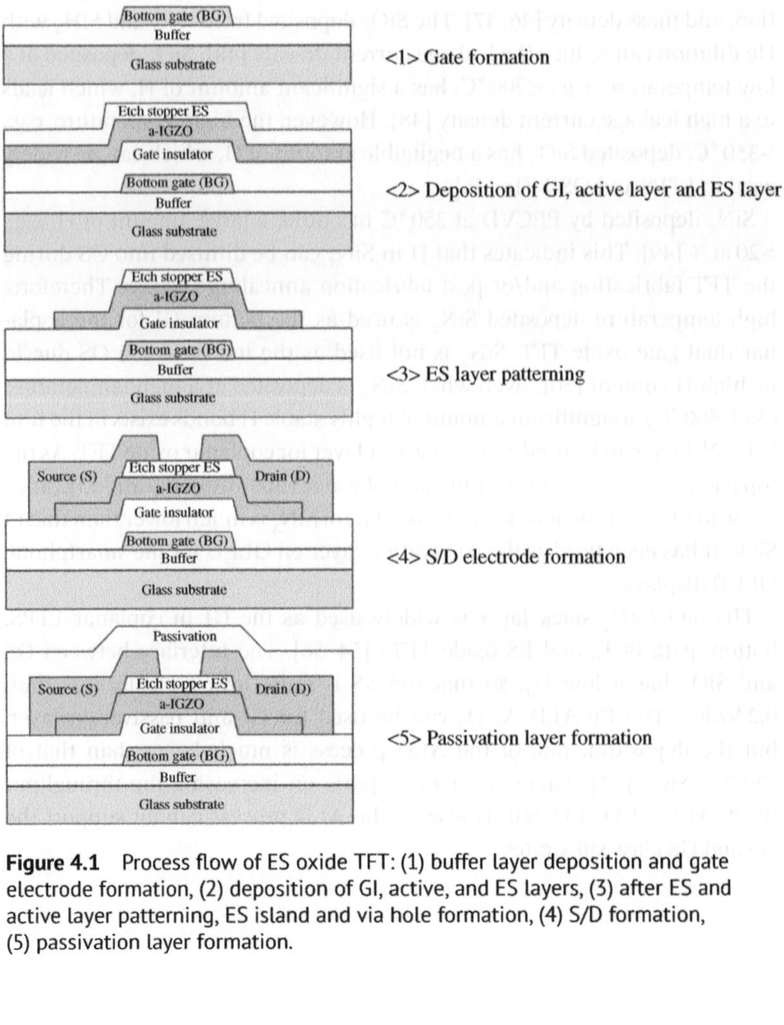

Figure 4.1 Process flow of ES oxide TFT: (1) buffer layer deposition and gate electrode formation, (2) deposition of GI, active, and ES layers, (3) after ES and active layer patterning, ES island and via hole formation, (4) S/D formation, (5) passivation layer formation.

Table 4.1 Performances of ES IGZO TFTs reported early stage of development.

V_{th} (V)	μ_{FE} (cm²/V s)	SS (V/dec.)	PBTS (V_{Stress}, Temp., ΔV_{th})	W/L (μm)	Year	References
2.2	26	0.214	20 V, –, < +1 V	25/55	2012	[65]
−0.10	27.7	0.11	20 V, 60 °C, +0.6 V	40/20	2013	[67]
−0.05	8.8	0.28	30 V, 60 °C +0.02 V	12/6.5	2014	[68]
0.9	15.3	0.23	20 V, 100 °C, +2.5 V	50/20	2014	[63]

ES and OS layer patterning, ES island and via hole formations, (4) S/D electrodes formation, and finally (5) passivation layer formation. The ES layer can shield the a-IGZO during the etching process of the S/D electrodes. It, however, introduces two disadvantages: (i) it requires an additional photo-mask step, which increases the manufacturing cost, and (ii) it has misalignment margins between gate and ES, and between ES and S/D, which results in a large gate-to-drain overlap capacitance [64]. This induces a large kick-back voltage, degrading the display performance [65, 66]. A summary of the ES IGZO TFTs reported in the early stage of oxide TFT research can be seen in Table 4.1. The self-aligned (SA) top gate, coplanar structure is getting more popular for display TFT backplanes because of the advantage of negligible parasitic capacitance between the gate and source/drain electrodes [68–70]. Therefore, it can reduce the resistance-capacitance (*RC*) delay for large-area OLED TVs, digital signages, and high-resolution displays. Note that the n^+ a-IGZO region between the source/drain contacts and a-IGZO channel should have very low resistance for oxide TFT backplanes in OLED displays [69, 70]. This low resistance is crucial for efficient current flow and overall performance of the TFTs. The ES structure can minimize the damage to the active layer during the fabrication process [67, 71, 72]. The ES structured oxide TFT is suitable for small-sized TFT products.

4.3.2 Back Channel Etch IGZO TFTs

The BCE structure is used for a-Si:H TFT LCD manufacturing [73]. The process is simple and the manufacturing cost is low. The a-IGZO TFT can form an ohmic contact with metals, such as Ti and Mo, without the need for a heavily doped n^+ layer [74]. Therefore, existing a-Si:H TFT production lines can be easily modified for BCE oxide TFT lines to reduce the initial capital investment. However, the OS is vulnerable to damage during the wet-etching process carried out for source/drain electrodes etch, e.g., the

$HNO_3 + H_3PO_4 + CH_3COOH$ solution, which leads to defects generation on the exposed oxide back surface [75, 76]. While source/drain electrodes can be patterned by dry etching, this can result in plasma damage on the active layer. To mitigate this, several techniques have been developed to avoid damaging the oxide semiconductor during the etching of the source/drain electrodes. For example, less acidic etchants, such as hydrogen peroxide (H_2O_2)-based solutions, are used instead of the commonly used nitric acid (HNO_3) based etchants for the a-IGZO back channel etch. Several groups have demonstrated BCE a-IGZO TFTs [77–80]. The BCE oxide TFT process is summarized in Figure 4.2, which includes steps of (1) bottom gate formation, (2) deposition and patterning of active and GI layer sequentially, (3) source/drain electrode formation, and finally (4) passivation layer formation. Although the BCE IGZO TFT has less paratactic capacitance compared to the ES TFT, it does have overlap capacitance between the gate and source/drain electrodes. A summary of the BCE structured IGZO TFTs reported in the early stage development can be seen in Table 4.2.

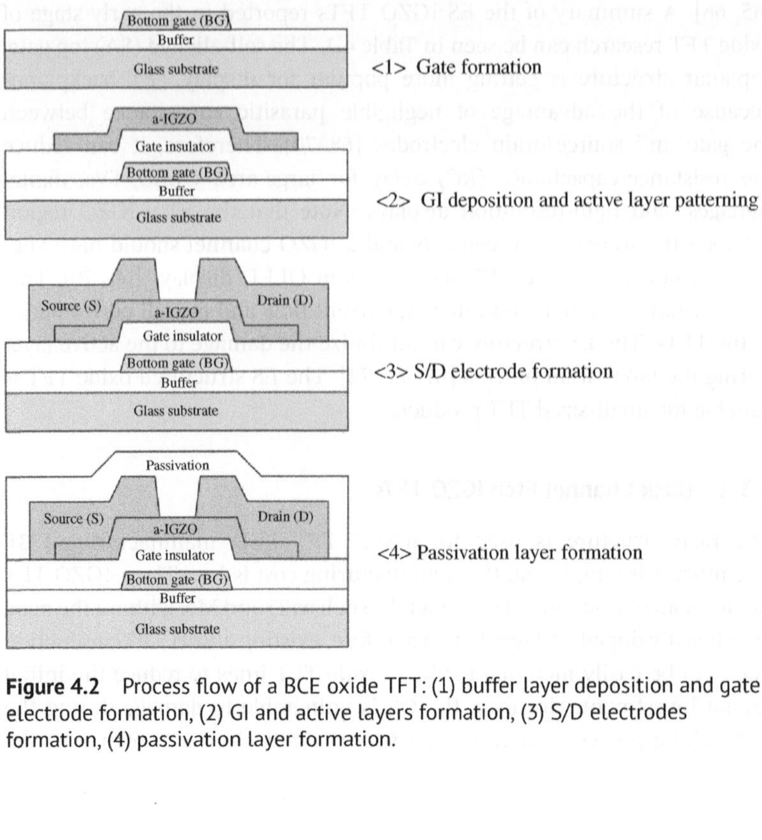

Figure 4.2 Process flow of a BCE oxide TFT: (1) buffer layer deposition and gate electrode formation, (2) GI and active layers formation, (3) S/D electrodes formation, (4) passivation layer formation.

Table 4.2 BCE IGZO TFT performances in early stage of research.

V_{th} (V)	μ_{FE} (cm²/V s)	SS (V/dec.)	BTS (V_{Stress}, Temp., ΔV_{th})	W/L (μm)	Year	References
0.1	16.9	0.16	20 V, –, +0.77 V	24/10	2010	[80]
7.0	10.2	0.52	20 V, 60 °C, +1.3 V	—	2012	[77]
−3.14	35.7	0.185	20 V, 60 °C, +3 V	2/2	2013	[78]
1.2	8.9	0.36	30 V, –, +1.21 V	—	2015	[79]

4.3.3 Coplanar IGZO TFTs

The coplanar IGZO TFT uses self-aligned patterning to ensure uniformity and the negligible kick-back voltage over the display area [34]. The fabrication process flow of the coplanar TFT is shown in Figure 4.3, which includes steps of (1) deposition of buffer and OS layers, (2) GI and gate electrode deposition after active layer patterning, (3) gate electrode patterning and self-aligned patterning of GI layer, (4) plasma doping on active layer for n^+ region, (5) interlayer deposition and via-holes formation, and (6) formation of source/drain electrodes. The gate-to-drain capacitance (C_{gd}) is negligible in the SA coplanar TFT structure [69, 70]. The offset region between the source/drain and the channel should have very low resistivity. For example, the conductivity of the a-IGZO increases significantly when it is exposed to H_2 plasma because H acts as a donor in it [81, 82]. Deposition of SiN_x on IGZO can also increase the conductivity from the H diffusion [83, 84]. In addition, F incorporation can make IGZO a n^+ layer [85]. The Ar/He plasma treatment can lower the resistivity of the OS film [50, 86]. All the above methods can be used to prepare the n^+ region in coplanar TFTs.

The coplanar structure has an additional advantage in making short-channel TFTs [87]. Currently, OLED TVs have coplanar TFT backplane that is effective in reducing the RC delay over a large-area screen [88, 89]. A summary of the coplanar IGZO TFTs reported in the early stage can be seen in Table 4.3. A comparison of the advantages and disadvantages among BCE, ES, and coplanar TFTs for display application is summarized in Table 4.4.

4.3.4 Dual-gate IGZO TFTs

A dual-gate (DG) TFT has two gate electrodes, the primary gate is the bottom gate (BG), which is often electrically isolated from the secondary gate (top gate). Two gate potentials can be separately controlled. The bottom

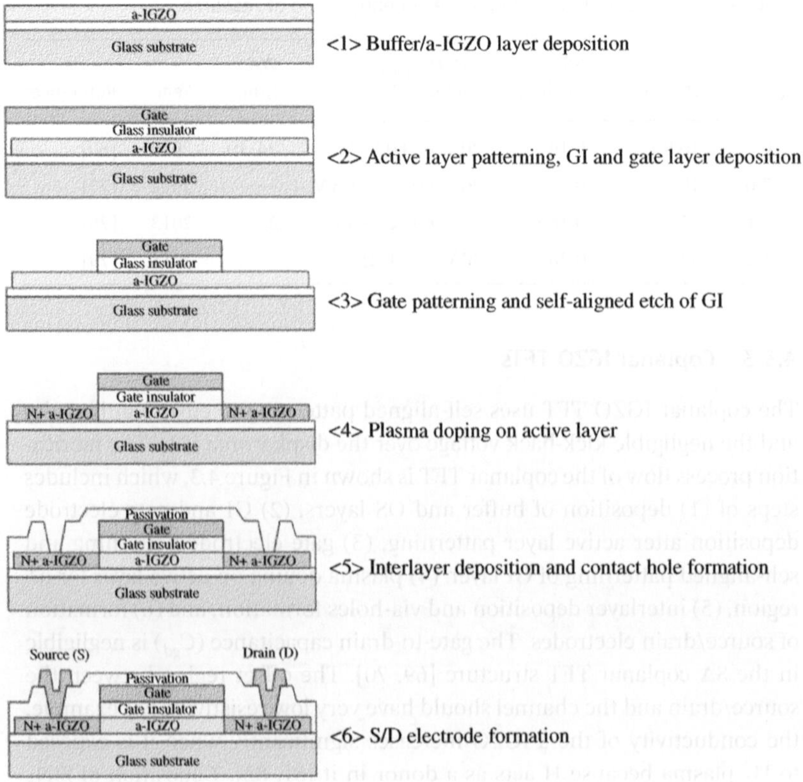

Figure 4.3 Fabrication process flow of a self-aligned, coplanar oxide TFT: (1) buffer layer (SiN_x/SiO_2) formation and active layer deposition, (2) active layer patterning, GI and gate electrode formation, (3) gate electrode and GI patterning using self-aligned process, (4) S/D doping by plasma treatment, (5) interlayer deposition and contact hole formation, and (6) S/D electrode formation.

Table 4.3 TFT performances of coplanar IGZO TFTs reported early stage of research.

V_{th} (V)	μ_{FE} (cm^2/V s)	SS (V/dec)	BTS (V_{Stress}, Temp., ΔV_{th})	W/L (μm)	Year	References
−0.6	22.2	0.15	20 V, –, +0.3 V	20/10	2013	[90]
—	15	—	20 V, –, +0.3 V	20/2	2013	[87]
0.55	12.1	0.10	30 V, 60 °C, +0.70 V	10/6	2014	[68]

Table 4.4 Advantages and disadvantages among BCE, ES, and coplanar TFTs in display applications.

Type	BCE	ES	Coplanar
Structure	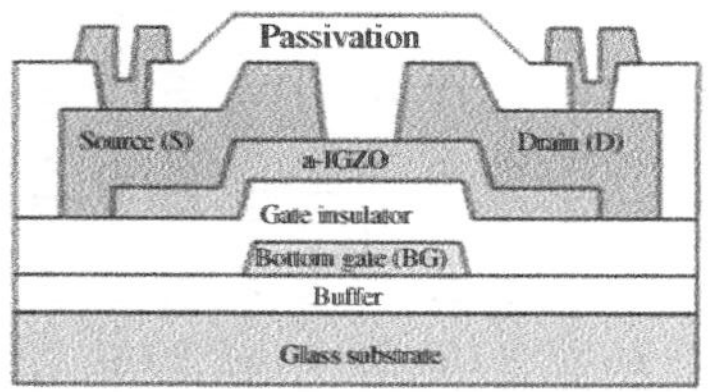	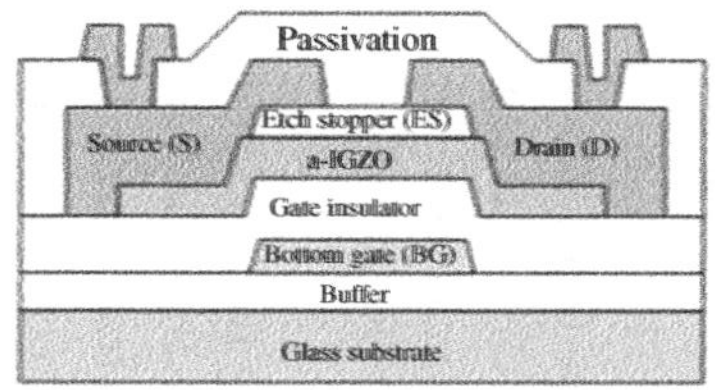	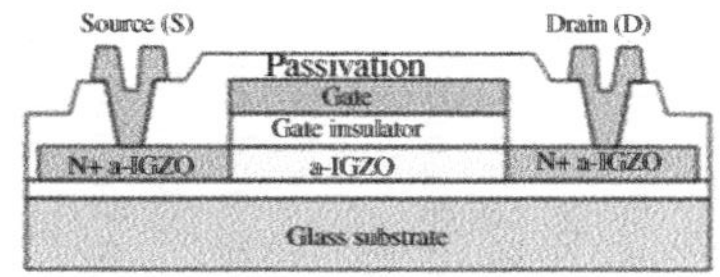
Advantage	• Fewer masks • Most compatible to a-Si TFT	• Protection of back channel • Better reliability	• Fewer masks • Minimum C_{gd}
Disadvantage	• Damage of back channel interface	• Need more mask • Large overlap Cap	• Damage of offset interface • Resistance of offset region

gate can be swept while the top gate voltage (V_{TG}) is kept at a constant value. On the other hand, the top gate can also be swept while the bottom gate voltage (V_{BG}) is kept at a constant value. Both gates could be electrically connected, which is the dual-gate (DG), dual-sweep (DS) mode. In this case, all active layers can have carriers accumulated, which is called the bulk-accumulation mode (BA) TFT. The DG, DS driving, BA mode TFT provides a high drain current (I_{DS}) due to the formation of two parallel carrier accumulation regions at the top and bottom interfaces, i.e., back and front channels [91, 92]. The back and front channels formed during DG, DS driving can be overlapped when the DOS is low enough in the gap of OS. This leads to high-performance TFT. For example, this results in the increase of I_{DS} by three to six times, reduction of *SS*, and low turn-on voltage (V_{ON}) of near 0 V compared to the single gate (SG) TFT in the saturation region [93–95]. The *SS* of the BA TFT is very low due to the fast filling-up of electrons in the gap states, as the Fermi level (E_F) moves fast towards the conduction band (E_C). Then, the OFF-state can be quickly switched to the ON-state. It is noted that DG-driving gives better V_{ON} uniformity compared to SG-driving [93–95]. The use of DG TFTs is reported for V_{th} control and I_{DS} enhancement.

The structure of DG TFT can be BCE, as shown in Figure 4.4, and coplanar, as shown in Figure 4.5. Note that there are overlaps between source/drain and bottom and top gate electrodes in the DG BCE structure. However, there is negligible overlap between the gate and source/drain in case of the DG coplanar structure. To realize OS-based circuits with high performance, n-type a-IGZO TFTs with different V_{th}'s are some cases required. The V_{th} can be adjusted by applying a constant TG potential while biasing the BG. Constant positive V_{TG} induces a negative shift of the transfer curve, thereby, enabling the depletion mode operation [52, 94].

The active layer thickness and DOS in the gap are important for high-performance BA TFT because the areal density of the gap states increases with OS thickness. However, the optimum thickness depends on the DOS in the gap of the OS. OS thickness can be larger when the total DOS in the gap is lower. The DG, DS driving can result in electron accumulation at the back and front interfaces as well as the whole active layer. Better TFT performance is attributed to the high gate drive from both sides. It is crucial to maintain the low density of interface states for both bottom and top GIs. When the active layer is thin and the density of states in the gap is sufficiently low, the two channels formed at the back and front interfaces can overlap, which results in the BA mode.

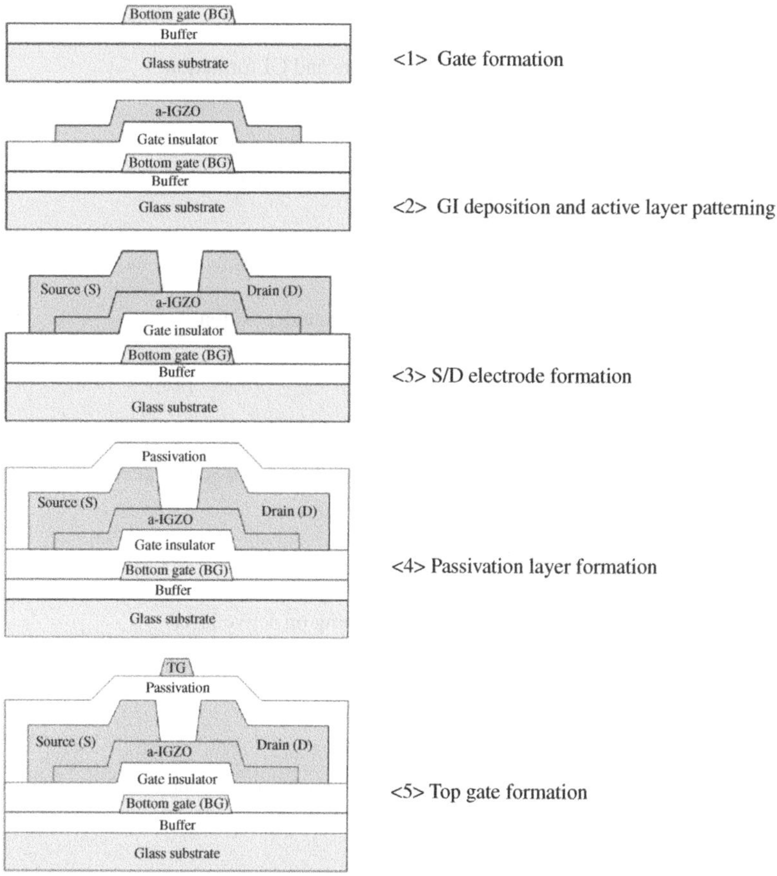

Figure 4.4 Fabrication process of a dual-gate, BCE oxide TFT: (1) buffer layer deposition and bottom gate formation, (2) GI and active layer formation sequentially, (3) S/D electrode formation, (4) passivation layer formation used as top GI, (5) top gate formation.

The effects of gate bias, temperature, light illumination, and mechanical strain stresses could be reduced by using BA mode [96]. When used in switching circuits, BA TFTs increase switching speeds and reduce power consumption. This improvement is possible because the driving voltage required for a BA TFT can be significantly lower than that needed for an SG TFT [92, 97, 98]. The performances of DG SA coplanar IGZO TFTs reported in the early stages of research can be seen in Table 4.5.

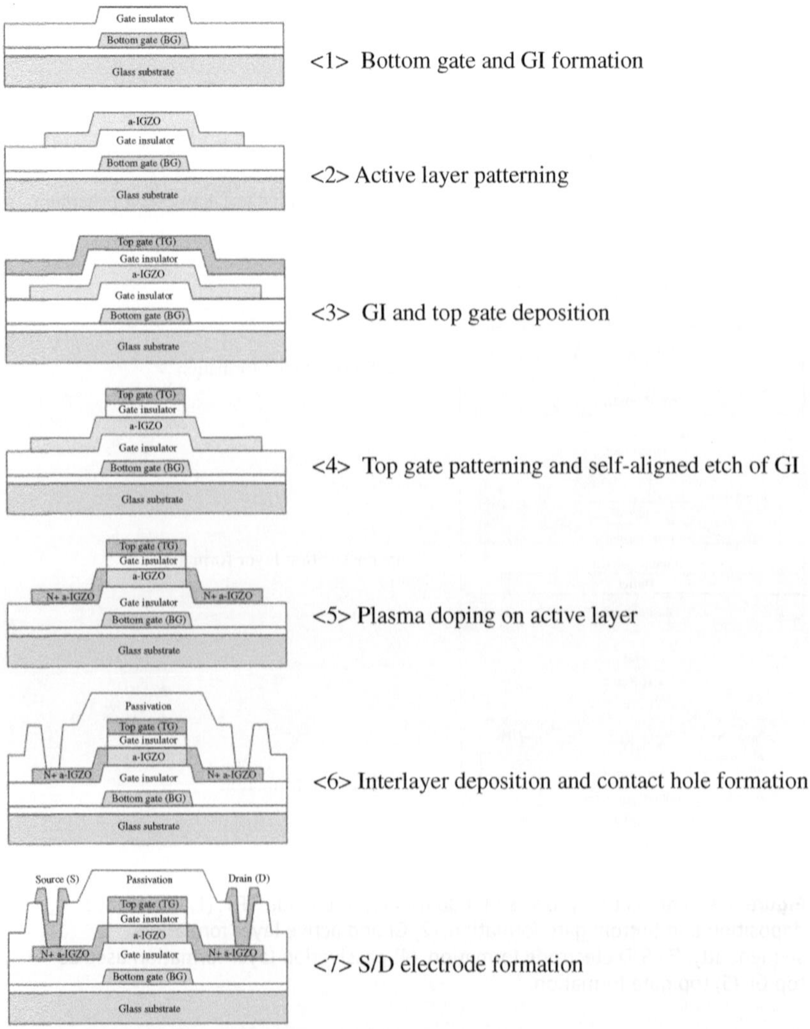

Figure 4.5 Fabrication process of dual-gate, coplanar oxide TFT: (1) buffer layer deposition and bottom gate electrode and bottom GI formation, (2) active layer formation, (3) top GI and top gate deposition, (4) top gate and top GI patterning, (5) plasma doping for n+ layer, (6) interlayer deposition and contact hole formation, (7) S/D formation.

Table 4.5 TFT performances of SA DG coplanar IGZO TFTs in early stage of research.

V_{th} (V)	μ_{FE} (cm^2/V s)	SS (V/dec.)	BTS (V_{Stress}, Temp., ΔV_{th})	W/L (μm)	Year	References
0.22	16.2	0.19	20 V, 60 °C, +2.7 V	40/10	2010	[99]
0.37	15.1	0.099	15 V, 80 °C, +4.5 V	60/10	2011	[100]

4.4 Flexible Substrate Technology for Oxide TFTs

4.4.1 Flexible Substrates

There is increasing interest in flexible electronics fabricated on plastic substrates. Since the displays fabricated on glass are brittle and heavy, they are insufficient for the needs of personal digital devices, which are required to be lightweight, thin, and bendable. The development of flexible displays can overcome these issues and provide design freedom for the shape and bending ability of flat panel displays. The advantages of flexible electronics include the reduction in weight and thickness, improved ruggedness, and the ability to flex, curve, conform, roll, and fold for better portability. The long-term vision for flexible electronics involves roll-to-roll processing, which allows for the reduction in capital investment and device cost because of the increase in throughput and reduction in materials used [101, 102].

The typical material used for the flexible substrate is solution-processed PI. There are many advantages of PI compared to other plastics, such as polyethylene terephthalate (PET) and polyethylene naphthalate (PEN), which include higher glass transition temperature, lower coefficient of thermal expansion (CTE), and higher process temperature. The PI allows the TFT fabrication process temperature as high as 450–500 °C, which is very important to achieve high performance and highly stable LTPS and oxide TFTs. Low CTE of less than 7 ppm/°C can help to do the process on carrier glass (3 ppm/°C) without deformation of plastic substrate.

The current technologies to manufacture electronic devices utilize glass and PI/carrier glass substrates. Modification to existing tools, development of new tools, new process recipes, and system integration are developed for the manufacturing of flexible electronic devices [103]. The development of flexible substrates and compatible debonding technology are widely used to fabricate flexible AMOLED displays. Several types of flexible substrates, including ultra-thin glass, plastic, and stainless-steel foils have been considered as possible substrates [104–107], but PI film is mostly used for flexible electronics.

4.4.1.1 Fabrication of Polyimide Substrate

Flexible substrates could be prepared by a laminating or coating method on a glass carrier, followed by fabricating TFT array and display devices. The PI substrate can be manufactured using a polymeric solution process on carrier glass and then, de-bonding it after display fabrication. Note that the concerns of the laminating type of substrate technology are (i) the difference in thermal expansion coefficients between plastic and glass and (ii) thermal damage of adhesive glue on the carrier glass. Currently, the popular flexible AMOLED is made by the PI coating process. The amber color PI varnish can be spin- or slit-coated on the carrier glass. The slit-coating method has been applied to G6.5 carrier glass for flexible AMOLED manufacturing. The process temperature can be 450–500 °C because the CTE of the PI is not much different from that of the carrier glass, i.e., is 4–7 ppm/°C vs. 3 ppm/°C. The manufacturing steps are glass cleaning, slit-coating of PI varnish, PI curing in a furnace chamber, and PECVD deposition of the gas barrier layer.

4.4.1.2 Laser Lift-off Detachment of Polyimide Substrate

The use of a glass carrier allows the overcoming of shrinkage-induced misalignment issues during many photolithographic steps in the TFT fabrication process. Finally, the samples are detached from the carrier glass using laser lift-off (LLO) process, as shown in Figure 4.6 [108]. Technologies of removing the de-bonding layer (DBL) from glass have also developed [109]. For example, flexible OLED display products are made using the LLO technology. The a-Si sacrificial layer is used for the easy detachment of the product from the carrier glass, as shown in Figure 4.6. The LLO technology is being used by display manufacturers because of its large-area compatibility [108]. In LLO, a long line of the laser beam is transmitted through the transparent carrier glass to the a-Si buffer layer between the PI layer and the carrier glass. The flexible substrate can be delaminated from the carrier glass, regardless of the strength of the adhesion force. The a-Si layer, which is evaporated by absorbing the ultraviolet (UV) laser, was proposed as the sacrificial layer to manipulate the adhesion of the PI film [105, 106]. It plays a crucial role in the LLO process. The a-Si can be easily evaporated by absorbing UV (for example, 308 nm), and thus can help to lift off the PI substrate from carrier glass.

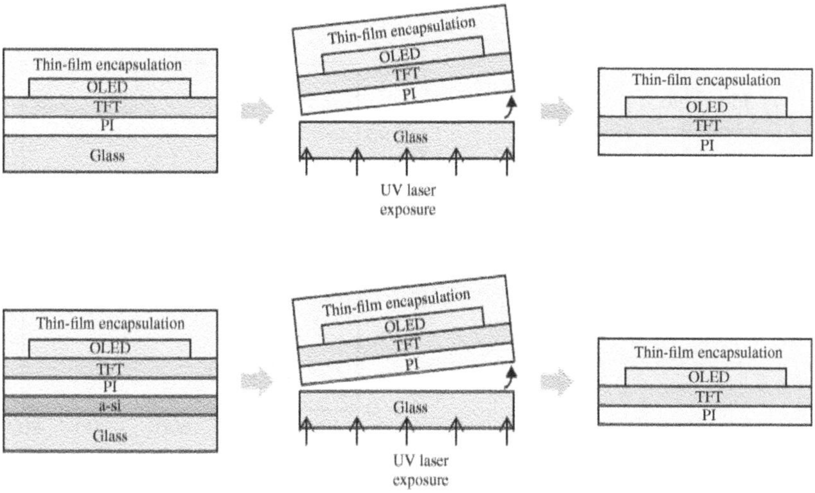

Figure 4.6 Process flow of laser lift-off method in detachment of OLED TFT on flexible PI.

4.4.1.3 Mechanical Detachment of Polyimide Substrate

A mechanical detach technology was developed using a carbon-nanotube (CNT):graphene-oxide (GO) mixture. The GO has a flake-like laminar structure to which the CNT needles are linked. The CNT:GO layer reduces the adhesion of the PI on carrier glass and thus can be used as a buffer for mechanical detachment, as shown in Figure 4.7. The solution-processed CNT:GO mixture can be spray-coated on the top of the carrier glass and then annealed before the PI solution coating, as shown in Figure 4.8.

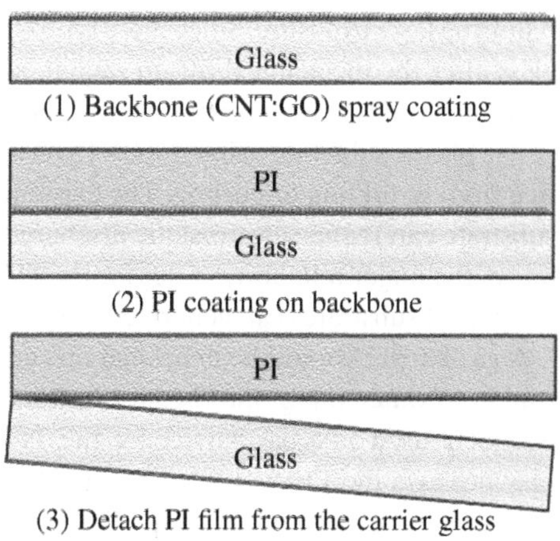

Figure 4.7 Process flow of non-laser detach method from carrier glass for flexible TFT backplanes: (1) very thin CNT:GO layer spray coated on carrier glass, (2) PI layer coating by slit or spin coating method on CNT:GO/carrier glass, (3) mechanical detachment of the PI substrate from the carrier glass after TFT backplane process.

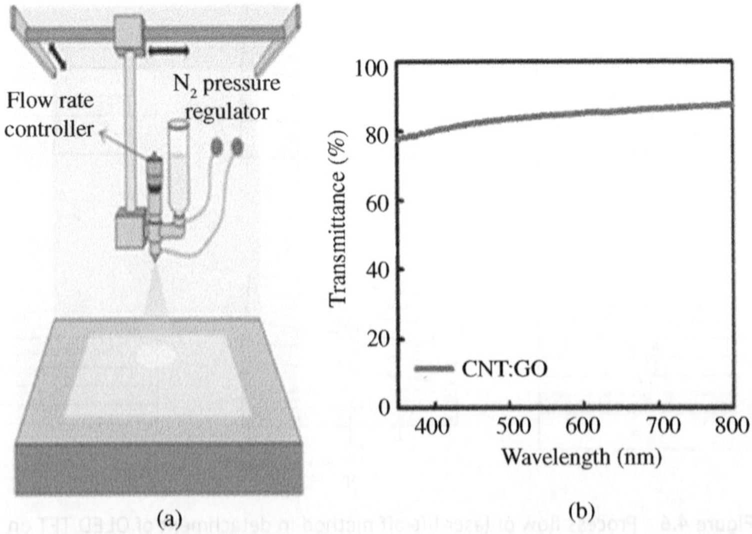

Figure 4.8 (a) Schematic view of a spray-coating machine used for deposition of CNT:GO mixture, (b) transmittances of a CNT: GO (1:1) layer with sheet resistance of ~10 kΩ/square.

After detachment from the carrier glass, the CNT:GO is embedded on the back side of the PI substrate, which is visible from the high-resolution scanning electron microscope (SEM) image. There is no CNT:GO residues left on the carrier glass.

The CNT layer on the back of the PI substrate can decrease the contact area with the carrier glass because of its hydrophobic properties. This makes it possible to lift off the PI layer from the carrier glass mechanically without requiring laser exposure. Since there is no CNT:GO residue left on the carrier glass after the mechanical lift-up step, it is a better DBL matter than other buffer materials, which remain on the glass [109, 110].

The plastic substrate using the CNT:GO lift-off process can be standalone, free from wrinkling or curling. The existence of CNT on the backside of PI substrate can reduce electrostatic discharge (ESD) damage. ESD damage is one of the major reliability issues for the TFTs on PI flexible substrate. Handling of PI substrates can generate ESD damage on it.

A gas barrier should be deposited on the PI substrate to block the penetration of water vapor and the absorption of O_2. The SiO_2/SiN_x multilayer is widely used as a gas barrier for the TFT backplane for the OLED display because the PECVD SiN_x is a good gas barrier. A gas barrier should be deposited on top of an AMOLED to protect the OLED and TFT from exposure to ambient air. In general, organic films are not good barriers for humidity and O_2. If needed, the organic layer is sandwiched between two SiN_x layers. For example, the SiN_x/organic film/SiN_x barrier structure has been widely used in flexible AMOLED displays because of its simple process. The organic layer is usually coated by the inkjet printing method.

4.4.2 Flexible IGZO TFTs on PI Substrate

The detailed fabrication process of a-IGZO TFTs has been explained earlier in this chapter and also in other literature reports [111–114]. The fabrication process for the ES TFTs on PI substrate is illustrated in Figure 4.9. Between the TFT devices and the PI substrate, a gas barrier consisting of a stack of alternating SiO_2 and SiN_x layers is inserted. The TFTs could be BCE, ES, or coplanar structures. The fabrication process includes steps of (1) CNT:GO spray coating and curing, (2) 10–20 μm PI substrate deposited by slit- or spin-coating followed by curing in a furnace, (3) deposition of a multilayer (SiN_x/SiO_2) gas barrier and an SiO_2 buffer layer on PI, (4) bottom gate formation, (5) deposition of GI, OS, and ES layers followed by patterning, (6) S/D electrodes formation, (7) passivation layer deposition and via holes formation, (8) detachment of the flexible TFT array with the PI substrate from the carrier glass.

The fabrication process of the SA, coplanar oxide TFT backplane on PI substrate is shown in Figure 4.10, which includes steps of (1) CNT:GO spray coating and curing, (2) PI deposition by slit-or spin-coating and subsequent curing, (3) PECVD of a multilayer (SiN_x/SiO_2) gas barrier and a SiO_2 buffer layer on PI, (4) OS deposition and island formation, (5) deposition of GI and metal layers, (6) gate electrode formation and SA etch of GI, (7) n^+ doping by plasma exposure, (8) interlayer deposition, (9) interlayer via hole formation, (10) S/D electrodes formation, (11) detachment of TFTs on PI from the carrier glass.

The electrical characteristics of IGZO TFTs on PI substrate can be changed under mechanical strain. The strain on the TFT can be obtained from the following Eq. (4.1) [115].

$$\varepsilon_{Surface} = \left(\frac{1}{R} \pm \frac{1}{R_O}\right) \cdot \left(\frac{d_S + d_f}{2}\right) \cdot \left(\frac{1 + 2\eta + \chi\eta^2}{1 + (1+\eta)\cdot(1+\eta\chi)}\right) \tag{4.1}$$

where, R_o is the radius of curvature of the structure before applying strain, R is banding radius, d_s is substrate thickness, d_f is TFT film thickness, E_f is Young's modulus of TFT film, E_s is Young's modulus of plastic substrate, $\eta = d_f/d_s$ and $\chi = E_f/E_s$.

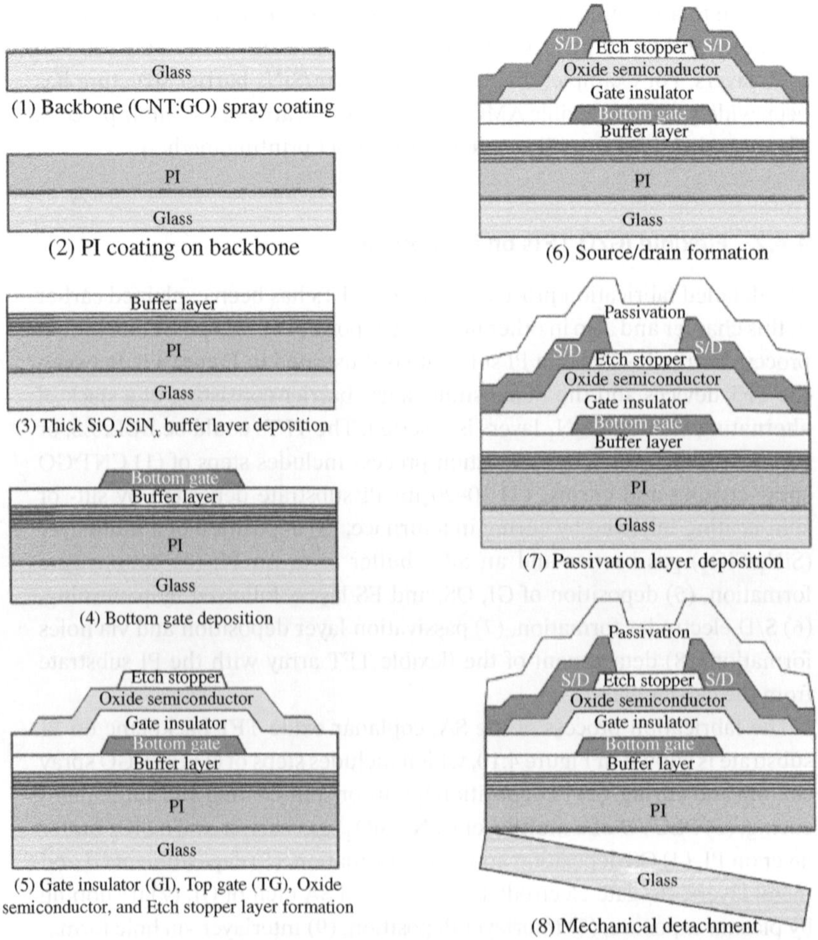

Figure 4.9 Fabrication process of a flexible ES oxide TFT backplane: (1) CNT:GO coating, (2) PI coating and curing, (3) deposition of $(SiN_x/SiO_2)_n$ gas barrier and SiO_2 buffer layers, (4) bottom gate formation; (5) depositions of GI, oxide semiconductor, and ES as well as ES island formation, (6) S/D formation, (7) passivation layer deposition and vis holes formation, (8) detachment from carrier glass.

Mechanical stability of the TFT device is a big issue for flexible electronic applications [116, 117]. Particularly, metal electrodes used for gate and S/D in flexible displays are brittle under mechanical stress [118–120]. In order to obtain highly flexible and bendable oxide TFTs, the mechanical strain can be reduced by using small Young's modulus material [121, 122], very thin substrate [123], metal nanowire (NW) [124], and neutral plane (NP)

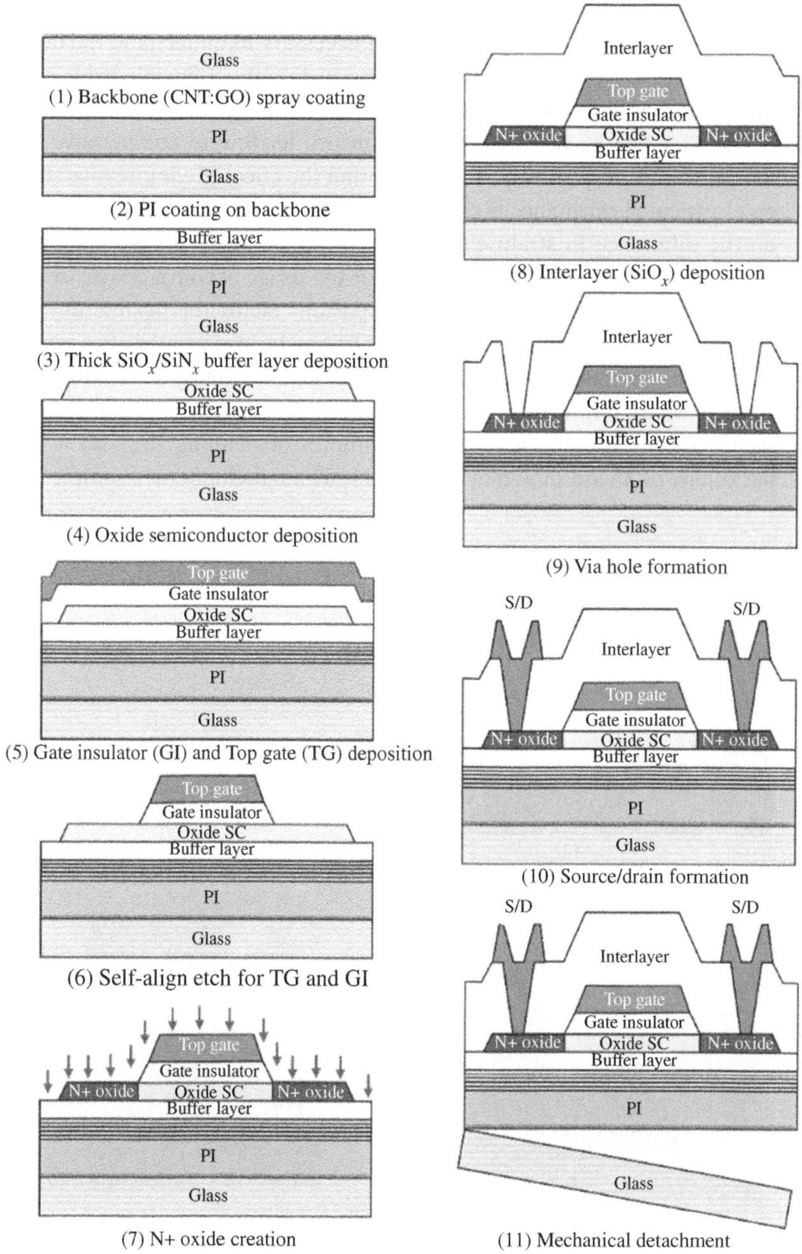

Figure 4.10 Fabrication process of the flexible coplanar oxide TFT backplane: (1) CNT:GO coating and curing, (2) PI coating and curing, (3) deposition of $(SiN_x/SiO_2)_n$ gas barrier and SiO_2 buffer layers, (4) oxide semiconductor island formation, (5) deposition of GI and top metal, (6) patterning of top metal and GI, (7) n^+ doping by plasma exposure, (8) interlayer deposition, (9) interlayer via hole formation, (10) S/D formation, (11) detachment from carrier glass.

structures [125–127]. The PI can be printed on top of the TFT backplane to make it a neutral plane device. It is still necessary to understand the origin and to solve the problem of the electrical instability of flexible oxide TFTs under mechanical strain.

The TFT can be bent inward or outward, leading to compressive and tensile strains, respectively. It is known that the effect of compressive strain on electrical performance is smaller than that of tensile strain [128]. Based on the difference in strain effect, most of the current foldable AMOLED displays in smartphones are designed for the usage of compressive, inward folding. Mechanical stability is an important factor for flexible devices. The electrical properties of the a-IGZO TFT can be investigated after various tensile or compressive bending cycles. For the feasibility of flexible electronic devices, rolling tests are carried out under the tensile or compressive strain condition using a cylinder. Examples of bending tests on a TFT backplane on PI are shown in Figure 4.11, which includes (a) a sample on a

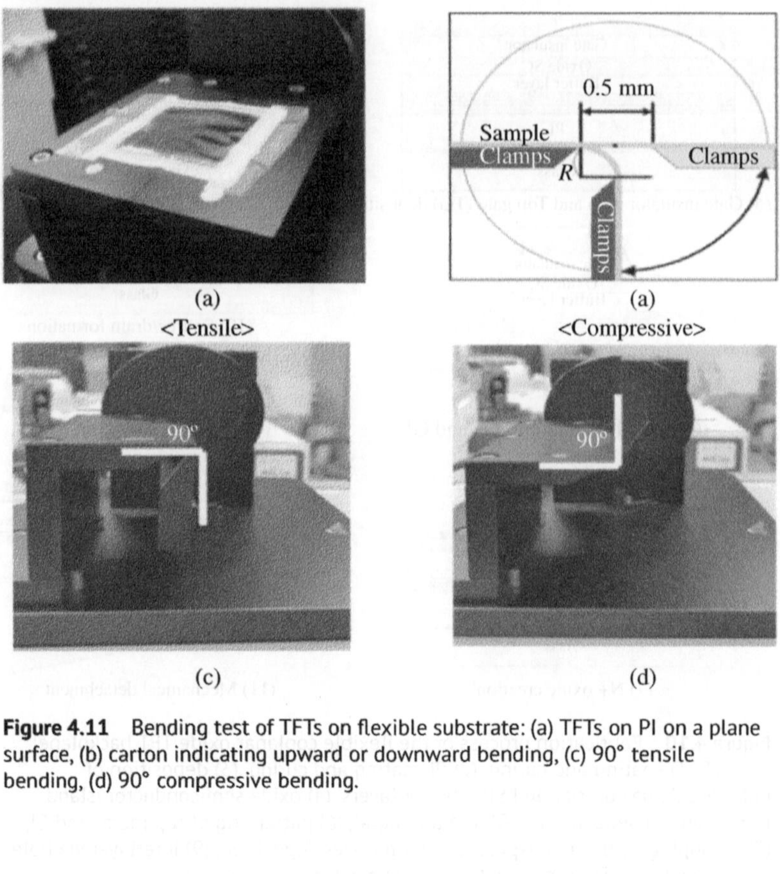

Figure 4.11 Bending test of TFTs on flexible substrate: (a) TFTs on PI on a plane surface, (b) carton indicating upward or downward bending, (c) 90° tensile bending, (d) 90° compressive bending.

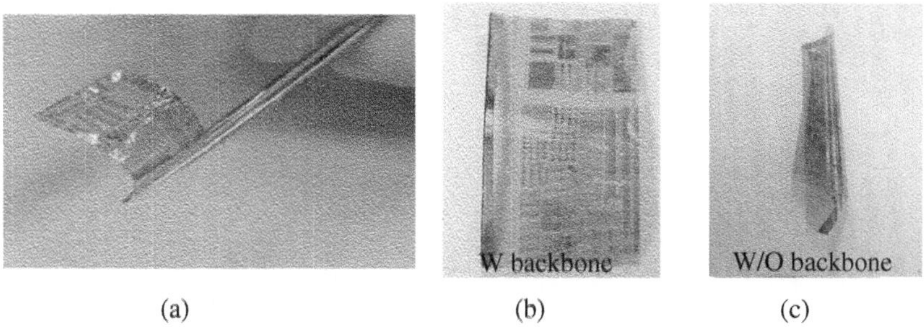

Figure 4.12 CNT:GO/PI/TFT samples: (a) free-standing TFT array on PI substrate with CNT:GO backbone, (b) free standing TFT array on PI substrate with CNT:GO backbone located on flat surface, (c) same sample with (b) but without the CNT:GO backbone.

flat surface, (b) bending directions, (c) tensile bending, and (d) compressive bending.

The backbone layer under the flexible PI substrate/TFT backplane is important for its self-standing capability. Figure 4.12 shows the TFT arrays on PI substrate: (a) free-standing TFT array on PI substrate with CNT: GO backbone, (b) free standing TFT array on PI substrate with CNT:GO backbone located on flat surface, and (c) same sample without the CNT: GO backbone. The existence of the CNT:GO layer prevents the flexible PI substrate from curling up.

4.4.3 Highly Flexible Oxide TFTs

Flexible electronic systems built on plastic substrates have been intensively studied because of the advantages of being bendable and rollable as well as being very thin and light [128–131]. The electrical characteristics of the a-IGZO TFT built on the flexible PI substrate are changed during and after mechanical bending. The mechanical stability of the TFT is a big issue for foldable and rollable AMOLED displays. Particularly, metals used for gate and S/D electrodes of flexible displays are brittle under bending stress [117, 132, 133]. In order to obtain highly flexible oxide TFTs, materials with small Young's modulus can have low strain during bending [134].

Other factors influence the mechanical properties of the oxide TFT's performance on the flexible substrate. For example, a very thin substrate [135], metal NWs [136], neutral plane structures [115], etc. can be used to improve the flexibility of the bendable devices. The size of nanowires [137–139], surface strain [140–142], anisotropy [136], surface roughness [139], crystalline defect [135, 143], etc. have been proved critical to the performance of flexible TFTs.

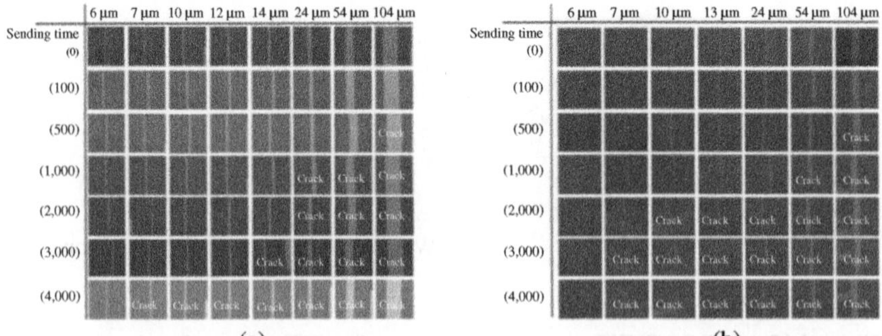

(a) (b)

Figure 4.13 Optical images of (a) Mo and (b) IGZO lines of 6 to 104 μm width after bending up to 4,000 times.

Figure 4.13 shows optical images of (a) Mo and (b) a-IGZO lines of 6–104 μm width after repeated bending at 90° for 4,000 cycles. The reduction of Mo and a-IGZO line width results in the reduction of crack formation. Cracks are observed on both Mo and a-IGZO lines wider than 7 μm, but no cracks are detected on 6 μm lines.

The I_{ON} drop in conventional TFT after mechanical bending can be contributed by the increase in series resistance, which comes from cracks generated in the Mo electrode. It was also observed that the TFT with meshed metal electrodes but non-patterned a-IGZO semiconductor layer showed the V_{th} shift toward the negative gate voltage (V_{GS}) direction after 4,000 bending cycles [141, 144, 145]. This mechanical strain effect is similar to the negative bias illumination stress (NBIS) effect, i.e., the generation of double ionized V_{o}^{2+} and/or single ionized V_{O}^{+} [146] as well as the donation of free electrons into the conduction band. Note that the oxygen vacancies generated by mechanical strain can be removed by thermal annealing, e.g., at 300 °C for one hour [147].

The TFT with the strip-patterned a-IGZO semiconductor layer exhibits stable performance even after many bending cycles. When the bending direction is parallel to the strip line, the cracks generated in the a-IGZO semiconductor do not propagate throughout the entire line width. Therefore, the strain-induced free carrier concentration in the TFT with the non-patterned a-IGZO is higher than that of the TFT with the strip-patterned a-IGZO.

The generation of cracks can contribute to the increase of resistance within the channel and at the S/D to a-IGZO contact. Under extreme bending conditions, cracks can extend extensively to destroy the TFT. The S/D metal determines the contact resistance and can diffuse into the oxide

Table 4.6 Young's moduli of materials used in oxide TFTs.

Material	Young's modulus (GPa)	References
Polyimide	2.5	[114]
IGZO	137	[148]
SiO_2	66	[149]
SiN_x	290	[149]
Mo	329	[114]
Ti	116	[150]
Ni	207	[148]
ITO	116	[151]
Al	68	[148]
Al_2O_3	370	[148]
ZrO_2	200	[150]

semiconductor layer. Therefore, the Young's moduli of films in the oxide TFTs, as shown in Table 4.6 [114, 148–151], are critical to the degradation of the TFT under mechanical bending conditions.

Various types of S/D electrodes are used for IGZO TFTs such as indium tin oxide (ITO) [107], indium zinc oxide (IZO) [152], Mo [153], Al [154], and Cu [155]. Although oxide-based electrodes, such as ITO and IZO are transparent, their large electron affinities show nonohmic behavior in the output characteristics [156]. The Cu electrode can form a good ohmic contact with a-IGZO [143]. However, the inter-diffusion and/or interfacial reaction between Cu and a-IGZO can lead to a change in device performance [157]. The Mo and Ti electrodes form good ohmic contacts with a-IGZO and therefore, give excellent device performance. To prevent the propagation of generated cracks in the conventional (not-patterned) Mo electrode under mechanical bending, a mesh-patterned design can be used. The resulting TFT showed robust S/D under mechanical strain, but it can have a negative shift of V_{th} and an increase of I_{ON} when the IGZO layer is not split. This is due to the generation of O vacancies. Therefore, by reducing the metal and semiconductor line widths as well as using mesh and strip designs, crack generation in the oxide TFT can be greatly reduced [114].

It is noted that the TFT in neutral plane can be made by the deposition of a PI layer on top of it, which will be strongly resistant to the strain

[115]. Splitting both active semiconductor and metal bus lines can improve the mechanical stability of the TFT. The split or mesh-patterned S/D electrode can also improve the TFT stability [114]. The TFT in the BA mode can be more stable against mechanical stress than the conventional single-gate TFT.

4.5 Solution Process Prepared Oxide TFTs

4.5.1 Precursors for Semiconductors and Gate Insulators

The precursors of OS and GI could be deposited by spin coating, spray pyrolysis, dip coating, and inkjet printing methods. The uniformity, coffee ring-free, bubble-free, material aggregation-free, high density, and low surface roughness should be considered for a good quality film. The performance of the film mainly depends on the quality of the precursor solution, including solubility, homogeneity, ionic strength (pH value), and method of deposition.

The solubility of an OS precursor in a solution depends on the type of chemical, such as chloride [158–165], nitrate [160, 161, 166–169], or acetate [168, 170–172] based metal and the corresponding solvent, such as methanol [159], 2-methoxyethanol (2ME) [158, 159, 163–165, 173], deionized water (DI) [162, 171], acetonitrile, ethylene-glycol (EG) [160, 174, 175], *N*,*N*-dimethylformamide (DMF) [176, 177], or 2-propanol [168]. The precursor should have good solubility and proper ionic strength. In addition, stabilizers, such as ammonium acetate, ethanolamine, or ammonium compounds, are added to the solution to enhance the solubility as well as to control the pH value. Sometimes solvents of different boiling points (BP's) are mixed, such as acetonitrile with EG, DMF with methanol. A low BP solvent helps to evaporate the organic components quickly to obtain a smooth film. On the contrary, a high BP solvent may improve the solubility of the solution, resulting in a good-quality film [176, 177].

The choice of the precursor and the solvent is important for a particular metal-oxide film because the process temperature is depended on them. The concentration of the precursor in the solution is critical to the film's uniformity and porosity. The low vapor pressure may meet challenges in film uniformity and contamination control. In addition, the decomposition of ligands can contribute to impurities in the final film. Maintaining the precise stoichiometry of the precursor is necessary in achieving the desired film properties. The deposition condition needs to be optimized for the high-quality material.

The chloride-based metal precursors with the 2ME and mixed solvent, such as acetonitrile and ethylene glycol, can be spin coated to prepare the good quality GI thin film [168, 178]. Since the BPs of acetonitrile and EG are 82 and 197.3 °C, respectively [175], during the curing process, the solvents are evaporated in two steps, which can provide a dense film with a proper M–O–M structure. The spray pyrolysis deposited GI film using the acetylacetonate-based metal precursor with the DMF/methanol mixed solvent can have good properties [176, 177]. Separately, the spin-coated oxide using the chloride-based precursor with the acetonitrile/EG and 2ME solvent showed promising semiconductor properties [158, 174, 175]. Another promising method of preparing the good oxide semiconductor thin film is the spray pyrolysis of the solution composed of acetate and nitrate-based metal precursor and 2ME solvent including a stabilizer, ammonium acetate, and ethanolamine [164, 168, 179–182].

Tables 4.7 and 4.8 show lists of precursors and solvents that have been demonstrated useful in preparing good quality OS and GI films respectively for oxide TFTs by the solution process.

4.5.2 Solution-Processed Oxide Semiconductors

Next-generation display demands TFTs with improved mobility, high I_{ON}/I_{OFF} ratio, reduced production cost, and uniform device performances over a large area. Note that LTPS on glass can be manufactured up to G6.5 glass ($1.5 \times 1.8\,m^2$), which is much smaller than G10.5 glass ($2.9 \times 3.4\,m^2$). Spray pyrolysis is a promising deposition method due to its low-cost and large-area, e.g., on G10.5 glass, process capability. Other types of solution-process methods, such as spin-coating, suffer from poor film uniformity over a large area.

Nowadays, most AMFPD backplanes are based on a-Si:H and poly-Si TFTs. The drawbacks of a-Si:H devices are low mobility ($<1\,cm^2/V\,s$) and poor stability after a long period of bias stress. Metal oxide semiconductors can substitute Si TFTs for display applications. The multicomponent metal-oxide semiconductors, such as IGZO, aluminum-indium zinc oxide (AIZO), LaZnO, F doped ZnO, zinc-tin oxide (ZTO), indium-zinc-tin oxide (IZTO), indium-zinc oxide (IZO), and IGZTO, are widely used for oxide TFTs [164, 178, 186, 188–191].

Superlattice and multi-stack heterojunction structures can be used in oxide devices to improve electrical performance. In a hetero-junction structure, mobility enhancement can be achieved by forming 2-dimensional electron gas (2DEG) near the interface of the junctions between two metal-oxide layers. The superlattice structure consists of a multilayer with

Table 4.7 Precursors and solvents used in solution process for oxide semiconductors.

OS	Precursors (chemical formula)	Solvent (stabilizer)	Deposition	References
In_2O_3	Indium(III) chloride ($InCl_3$)	2ME (ethanolamine)	Spin coating	[158]
In_2O_3	Indium(III) nitrate hydrate ($In(NO_3)_3 \cdot xH_2O$)	2ME, DI	Spin coating Spray pyrolysis	[166]
In_2O_3	Indium(III) acetate ($In(C_2H_3O_2)_3$)	2ME, DI	Spin coating Spray pyrolysis	[170]
ZnO	Zinc acetate dihydrate ($Zn\,(CH_3COO)_2 \cdot 2H_2O$)	2ME (ammonium acetate)	Spray pyrolysis	[179]
ZnO	Zinc chloride ($ZnCl_2$)	2ME, methanol, DI	Spin coating	[159]
ZnO	Zinc nitrate hexahydrate ($Zn\,(NO_3)_2 \cdot H_2O$)	Ethanol/DI ammonium soln	Dip coating	[169]
IGO	Indium(III) chloride ($InCl_3$) Gallium(III) nitrate hydrate ($Ga(NO_3)_3 \cdot xH_2O$)	Acetonitrile, EG, 2ME	Spin coating Spray pyrolysis	[160, 161]
IZO	Indium(III) chloride ($InCl_3$) Zinc acetate dihydrate ($Zn\,(CH_3COO)_2 \cdot 2H_2O$)	DI	Spin coating	[162]
IGZO	Indium(III) chloride ($InCl_3$) Zinc acetate dihydrate ($Zn\,(CH_3COO)_2 \cdot 2H_2O$) Gallium(III) nitrate hydrate ($Ga(NO_3)_3 \cdot xH_2O$)	2ME (ethanolamine, ammonium acetate)	Spin coating Spray pyrolysis	[163–165]

LiZnO	Zinc acetate dihydrate ($Zn(CH_3COO)_2 \cdot 2H_2O$) Lithium chloride hydrate ($LiCl \cdot xH_2O$)	2-Propanol (ethanolamine) 2ME (ammonium acetate)	Spin coating Spray pyrolysis	[168, 173]
GdZnQ	Gadolinium(III) nitrate hexahydrate $Gd(NO_3)_3 \cdot 6H_2O$ Zinc acetate dihydrate Zn ($CH_3COO)_2 \cdot 2H_2O$	2ME (ammonium acetate)	Spray pyrolysis	[168]
ZTO	Zinc chloride ($ZnCl_2$), Tin(II) chloride ($SnCl_2$)	Acetonitrile, EG	Spin coating	[174]
MgIZO	Zinc acetate dihydrate (Zn ($CH_3COO)_2 \cdot 2H_2O$) Indium(III) nitrate hydrate ($In(NO_3)_3 \cdot xH_2O$) Magnesium nitrate hydrate ($Mg(NO_3)_2 \cdot xH_2O$)	DI	Spin coating	[171]
CuO	Copper acetate monohydrate ($Cu(COOCH_3)_2 \cdot H_2O$)	DMF (ethanolamine)	Spin coating	[172]
SnO_2	Tin(II) chloride ($SnCl_2$)	Acetonitrile, ethylene glycol	Spin coating	[175]

Table 4.8 Precursors and solvents used in solution process for GIs.

GI	Precursors (chemical formula)	Solvent	Deposition	References
Al_2O_3	Aluminum chloride ($AlCl_3$)	Acetonitrile, EG	Spin coating	[183]
ZrO_2	Zirconyl chloride octahydrate ($ZrOCl_2{\cdot}8H_2O$)	Acetonitrile, EG	Spin coating	[168]
Y_2O_3	Yttrium(III) nitrate hexahydrate ($Y(NO_3)_3{\cdot}6H_2O$)	DI	Spin coating	[184]
HfO_2	Hafnium(IV) chloride ($HfCl_4$)	Acetonitrile, EG	Spin coating	[185]
Gd_2O_3	Gadolinium nitrate hexahydrate ($Gd(NO_3)_3{\cdot}6H_2O$	2ME, D I	Spin coating	[186]
LaAlO	Lanthanum nitrate hexahydrate ($La(NO_3)_3{\cdot}6H_2O$) Aluminum nitrate nonahydrate ($Al(NO_3)_3{\cdot}9H_2O$)	2ME	Spin coating	[187]
ZrAlO	Zirconium acetylacetonate ($Zr(C_5H_7O_2)_4$) Aluminum acetylacetonate ($Al(C_5H_7O_2)_3$)	2ME, DMF, Methanol	Spray pyrolysis	[176]

many interfaces. The In_2O_3/ZnO heterojunction TFT shows a high mobility of 45 cm^2/V s, resulting from the formation of 2D-like electron gas [192]. Note that, the shift in threshold voltage is reduced by the existence of the strong Al—O bond when Al_2O_3 is the top passivation layer [193, 194].

When the oxide TFT is exposed to air, desorption/adsorption of O_2 and water molecules may deteriorate the TFT, e.g., increasing hysteresis in the transfer characteristics with time, and it increases with increasing gate voltage [179, 195]. To shield the oxide TFTs from ambient air, several kinds of passivation layers, such as Al_2O_3, SiO_2, and Y_2O_3, have been studied [183, 193, 196, 197]. For long-term environmental stability, Y_2O_3 is a good passivation layer because of its large metal–oxygen bond dissociation energy of 714.1 ± 10.2 kJ/mol. It can improve the bias stability by decreasing deep-level defects in IGZO [198]. The solution-processed Y_2O_3 passivation has been used on oxide TFTs [183, 197, 199]. Y can diffuse into OS for a certain depth and decrease the oxygen vacancies.

4.5.3 Solution Process Prepared High-*k* Gate Dielectrics

Oxide TFTs typically use SiO_2 and SiN_x/SiO_2 stack as the GI for products. They are often operated with high V_{GS} and need high power consumption

[56, 200]. To reduce the operating voltage when used as a portable device, the SiO_2 thickness has to be reduced, which leads to the increase of the gate leakage current. To overcome the issue, the high-*k* dielectric material can be used as the GI. The oxide TFT with a high-*k* GI can have a high mobility and be operated at a low voltage, which is suitable for low-power applications [160, 201, 202]. The common high-*k* oxide gate dielectrics in oxide TFTs are zirconium oxide (ZrO_2), aluminum oxide (Al_2O_3), aluminum titanium dioxide ($AlTiO_2$), yttrium oxide (Y_2O_3), lanthanum-doped zirconium oxide (LaZrO) and HfO_2 [176, 184, 185, 198, 203, 204]. The high-*k* GI can be composed of more than one type of material, such as a mixture of lanthanum aluminum oxide (LAO) and zirconium aluminum oxide (ZAO), which remains amorphous with a smooth surface. The low surface roughness of the GI leads to fewer defects at the channel interface and therefore, resulting in the low *SS* in the TFT [176, 187, 205, 206]. However, since SiO_2 has many superior material properties compared with high-*k* dielectrics, such as the large band gap of 9 eV, large valence-band offset (VB) of 4.4 eV, and large conduction-band (CB) offset of 3.5 eV, high crystallization temperature, and excellent thermal stability, it is still a popular GI for oxide TFTs [42, 207, 208]. However, the oxide TFT has to be operated with a large V_{GS}, leading to high power consumption [206, 209]. Table 4.9 lists dielectric properties of SiO_2 and some high-*k* dielectrics used in oxide TFTs.

The solution-processed ZrO_2, HfO_2, and Al_2O_3 are the most popular high-*k* dielectric materials for their high dielectric permittivity, thermal stability, and low gate leakage current (I_G) contributed by their high band gaps and large conduction band offsets with OS. The band gap decreases

Table 4.9 Dielectric properties of high-*k* GI materials. Band offset values are with Si.

Material	*k*	Bando gap (eV)	CB offset (eV)	VB offset (eV)
SiO_2	3.9	9.0	3.5	4.4
Al_2O_3	9.0	8.8	2.8	4.9
Ta_2O_5	22	4.4	0.35	3.0
ZrO_2	25	5.8	1.5	3.3
HfO_2	25	5.8	1.4	3.4
La_2O_3	30	6.0	2.3	3.6
TiO_2	80	6.0	0.0	5.0
$SrTiO_3$	3000	3.2	0.0	2.2

gradually with the increase of dielectric permittivity. Among these materials, ZrO_2 is regarded as the most promising high-k gate dielectric due to its large band gap energy of 5.8 eV and large k value of ~23. Note that, it has a relatively large valence band offset (3.3 eV) with Si. However, its crystallization temperature is relatively low, e.g., about 600 °C. This problem can be solved by using ternary oxide high-k dielectrics, such as hafnium lanthanum oxide (HLO), lanthanum aluminum oxide (LAO), and lanthanum zirconium oxide (LZO) [204, 210–213]. The ZAO can be a good GI due to its amorphous nature and lower surface roughness. This contributes to the low interface state density and thus small *SS* [176, 177, 214].

The sol-gel deposition process starts with a solution of dissolved chemicals or colloidal particles which forms an integrated network of gel from hydrolysis and condensation reactions [215]. The sol-gel coating method can be used to deposit a single- or multi-component metal-oxide films. There are many advantages in the sol-gel coating process [216–218]. The chemical reactants of the process can be conveniently purified by distillation and crystallization. Since all starting materials are mixed at the molecular level in the solution, a high degree of homogeneity can be expected. Organic or inorganic salts can be added to adjust the microstructure or to improve the structural, optical, and electrical properties of oxide films.

4.5.4 Spin-coating Prepared IGZO TFTs

The spin-coating method has been used to deposit thin films for several decades. The typical process includes depositing a small amount of a resin fluid at the center of the substrate followed by spinning the substrate [219]. The film thickness is dependent on two parameters: fluid viscosity and spin speed. The thickness of the film increases with the increase of the solution's viscosity and the decrease of the spin speed. However, these parameters can affect the uniformity of the thin film. Multiple coating of a film may be required to prepare a device. Table 4.10 shows the electrical properties of some BCE structured IGZO TFTs prepared by spin-coating.

In the spin-coated IGZO films, the electron concentration decreases with increasing Ga/Zn ratio. This results in the decrease in Hall mobility and increase in resistivity [217, 218, 222, 226]. With increasing Ga concentration in IGZO, M–O density increases, and V_0 and OH concentrations decrease. On the other hand, with increasing In concentration, electron concentration and Hall mobility increase, and thus resistivity decreases. Thermal annealing of the spin-coated IGZO TFT increases its mobility and decreases *SS* by increasing M—O bonds and thus reducing O-related defects [215, 224].

Table 4.10 Performance summary of BCE structured IGZO TFTs fabricated by spin coating.

V_{th} (V)	μ_{FE} (cm^2/V s)	SS (V/dec.)	I_{ON}/I_{OFF}	Ratio (In:Ga:Zn)	Year	References
7.8	0.96	1.39	10^6	1:1:2	2009	[220]
7.8	0.96	1.39	10^6	3:0.65:2	2019	[221]
15.84	0.56	2.81	10^6	1:1:2	2010	[222]
3.52	0.90	1.16	10^6	3:1:2		
−5.09	1.25	1.05	10^6	5:1:2		
—	2	—	10^5	2:1:1	2013	[223]
—	0.2 (μ_{SAT})	—	10^5	63:10:27	2013	[224]
—	1.13 (μ_{SAT})	2.5	—	4:1:5	2010	[215]
—	5.8	0.28	10^7	3:1:1	2010	[216]
4.62	0.065 (μ_{SAT})	0.47	10^6	5:1:2	2010	[217]
—	5.43 (μ_{SAT})	—	10^7	80:10:10	2010	[218]
11.5	1.73	0.32	10^7	62:5:23	2009	[225]
—	0.72 (μ_{SAT})	0.68	10^6	68:10:22	2009	[219]
1.12	2.11	0.24	10^6	0.51:0.15:0.34	2023	[163]
−0.48	1.11	0.61	10^6	0.68:0.1:0.22	2022	[226]
3.70	1.27	0.65	10^6	—	2022	[227]
1.39	3.11	0.20	10^4	1:1:1	2023	[228]
0.17	12.67	0.47	—	4:1:4	2023	[229]

Chemicals, such as acetylacetone (AA) and trifluoro-acetylacetone (TFAA) are used to accelerate the reaction at low temperatures to achieve better performance [227]. The TFT inverters were developed by using spin-coated IGZO TFTs [228]. ZrO_2 high-*k* GI was also used in spin-coated IGZO TFTs [229].

4.5.5 Spray-pyrolysis Prepared IGZO TFTs

Spray pyrolysis is a promising technique to synthesize various high-purity homogeneous ceramic powders [230]. Figure 4.14 exhibits schematic views of (a) a spray-pyrolysis system and (b) the nucleation process of the metal-oxide film formation. The apparatus consists of an atomizer for the solution and a substrate. An aqueous solution containing soluble constituents of the desired compound is sprayed onto a heated substrate. Every

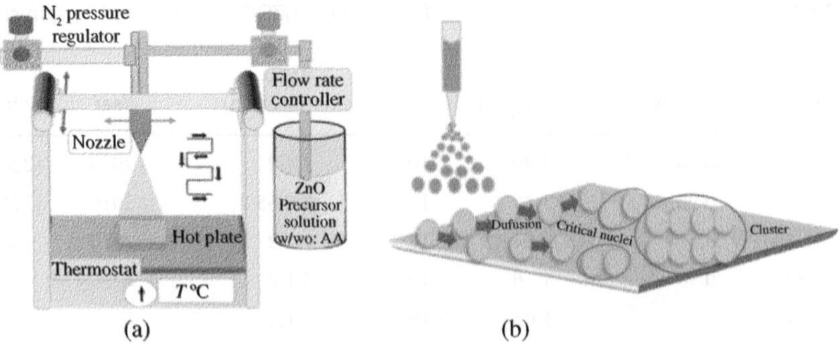

Figure 4.14 Schematic view of (a) a spray-pyrolysis coating system used for depositing metal-oxide films, (b) the nucleation process at the time of pyrolysis.

sprayed droplet reaching the substrate undergoes pyrolysis decomposition to form a large number of small, e.g., <100 nm, amorphous particles, individual crystallites, and/or a cluster of crystallites on the substrate surface homogeneously.

Deposition parameters, such as the volume of solution sprayed, substrate temperature, solution flow rate, and distance between the substrate and the spray nozzle, have to be optimized to get a good quality film. Hydrolysis and pyrolysis are the main chemical reactions involved in this process. In principle, the spray-pyrolysis process is simple, cost-effective, and suitable for industrial application [231].

The first a-IGZO TFT with the BCE structure fabricated using spray-coating was reported in 2015 [232]. The spray-coated a-IGZO exhibits thickness controllability, dense morphology, high electron mobility, and good electron transport properties comparable to those of the sputtered a-IGZO. In 2016, the same group reported the spray-coated all-oxide BCE a-IGZO TFT including the ITO gate and source/drain electrodes [233]. In 2018, an ultrasonic spray-coated BCE a-IGZO TFT was reported and its electrical characteristics are competitive to the sputter-deposited counterpart [234–236].

The BCE a-IGZO TFT has a significant parasitic capacitance from the overlap of the gate and source/drain electrodes, which is a challenging issue for the TFTs by spray pyrolysis. The first SA, coplanar a-IGZO TFT by spray-coating was reported in 2022 [164]. In this study, the n^+ a-IGZO was formed by NF_3 plasma treatment. Later, in 2023, the same group reported a spray coated coplanar a-IGZO TFT on a polyimide substrate for the low-cost flexible electronics [45].

Subsequently, the as-grown c-IGZO TFT was reported in 2023. It exhibited high mobility and excellent stability, better than the a-IGZO TFT

Table 4.11 Device properties of spray-pyrolysis prepared oxide TFTs of various structures.

Device structure	V_{th} (V)	μ_{FE} (cm^2/V s)	SS (V/dec.)	$I_{ON/IOFF}$	Ratio (In:Ga:Zn)	Year	References
BCE	0.05	19.10	—	$\sim10^5$	1.0:1.1:0.29	2015	[232]
BCE	−0.20	5.30	—	$\sim10^5$	3:1:1	2016	[233]
BCE	1.30	15.10	0.266	$\sim10^5$	1:1:1	2018	[234]
BCE	−7.20	1.74	—	$\sim10^6$	4:1:1	2020	[235]
BCE	2.70	9.04	0.099	$\sim10^7$	1:1:1	2021	[236]
Coplanar	−1.63	18.36	0.313	$\sim10^8$	1:1:1	2022	[164]
Coplanar	−0.80	40.95	0.180	$\sim10^8$	1:1:1	2023	[45]
Coplanar	−1.00	39.60	0.210	$\sim10^8$	1:1:1	2023	[237]
BCE	0.30	60.17	0.090	$\sim10^9$	1:1:1	2023	[165]
Coplanar	Synaptic TFT				1:1:1	2023	[238]

performance by spray coating [45, 165, 237]. IGZO was nucleated directly on the substrate and the grains grow three-dimensionally through repeated spray-coating steps. IGZO grains grow continuously until they collide to form grain boundaries. The c-IGZO by spray-pyrolysis exhibits a structure of highly oriented, *c*-axis-aligned crystal perpendicular to the substrate with a large mass density of 6.73 g/cm^3 without any disordered incubation layer.

Display manufacturing shifts from glass to flexible substrate and the cost should be low enough to compete in the market. One big advantage of spay-pyrolysis is the large-area deposition capability. Amorphous and polycrystalline oxide TFTs can be made on the PI substrate by spray-pyrolysis. Note that the spray-pyrolysis deposited amorphous $In_{0.5}Ga_{0.5}O$ (a-IGO) thin film, e.g., at ≥330 °C, is uniform and bubble-free [161]. The 370 °C deposited a-IGO thin film contains 5–8% nanocrystalline In_2O_3. The Leiden frost-dominated boiling (LFDB) grown a-IGO thin film exhibits good quality between 330 and 370 °C. The a-IGO TFT prepared at 330 °C showed good electrical characteristics, but the 290 °C prepared TFT showed poor characteristics. The TFT performance is related to the film density, V_o, and OH-related defects. Table 4.11 summarizes the device properties of spray-pyrolysis prepared TFTs of various structures.

4.5.6 Inkjet Printed Oxide TFTs

One of the big advantages of the solution process is that it can be carried out by inkjet printing. The IGZO layer can be directly printed on the channel

Table 4.12 Performance of inkjet printed IGZO TFTs reported in the literatures.

OS	V_{th} (V)	μ_{FE} (cm^2/V s)	SS (V/dec.)	$I_{ON/IOFF}$	Ratio (In:Ga:Zn)	Year	References
IGZO	−15.0	10.5	2	~10^5	2:1:1	2015	[239]
IGZO	8.7	0.071	—	~10^4	2:1:7	2019	[240]
IGZO	7.9	6.42 (μ_{SAT})	4.57	~10^5	72.5:7.5:20	2016	[241]
IGZO	0.0	20.6	0.19	~10^7	72.5:7.5:20	2013	[242]
IGZO	1.72	30.7	0.14	~10^6	6.8:1.0:2.2	2019	[243]
IGZO	6.2	0.03 (μ_{SAT})	1.50	~10^4	12.8:10.2:23.2	2009	[244]
IGZO	—	2.45	—	~10^5	57:12:31	2012	[245]
IGZO	—	6.2 (μ_{SAT})	—	—	6:1:2	2017	[246]
IGZO	2 ~ 4	2.6 ~ 3.3	—	~10^8	6:1:2	2018	[247]
In_2O_3/IGZO	11.0	3.9	1.5	~10^6	2:1:1	2019	[248]
In_2O_3/IGZO	0.5	18.7	0.53	~10^8	—	2021	[249]

region of the TFT, which saves the step of patterning the active layer [239, 240]. However, the performance of the inkjet printed IGZO TFT is not as good as that prepared by sputter deposition [241]. Note that the mobility of the inkjet printed IGZO TFT can be improved to 20.6 cm^2/V s accompanied with the improvement of *SS*, when a high-*k* dielectric of HfO_2 or Al_2O_3 is used as the GI [242, 243]. The inkjet printed IGZO can be crystallized at >700 °C, which is similar to the sputtered IGZO [244].

The performance of the oxide TFT, such as the mobility, can be improved when the bottom-gate, bottom-contact structure is used for inkjet printing [245]. Organic chemicals, such as Polyvinylpyrrolidone (PVP), doped IGZO can be prepared by inkjet printing [246]. The surface of GI can be exposed to UV/O_3 or treated with hexamethyldisilazane (HMDS) or octadecyltrichlorosilane (OTS) to make it hydrophilic for better wetting with an inkjet printed IGZO solution [247]. The composition of the IGZO can be easily adjusted. Also, heterojunction TFTs with multilayer semiconductors can be prepared by inkjet printing [248]. In general, the performance and stability of the inkjet printed oxide TFT are inferior to those of the sputtering or spray-pyrolysis prepared TFT. Table 4.12 summarizes the performance of the inkjet printed IGZO TFTs.

4.6 Summary

Oxide TFT backplanes have been used in AMOLED TVs and will be used in the future for G8.6 glass substrates for IT products. Reactive sputtering is the mainstream technology for OS deposition. The electrical properties of the oxide TFT, such as field effect mobility, threshold voltage, subthreshold swing, and bias stability, are strongly dependent on process parameters, such as Ar/O_2 partial pressure, substrate temperature, sputtering power, etc. The carrier concentration is also sensitive to the deposition condition so that the precise control of V_{th} is very difficult.

For large-size OLED TVs, the coplanar structured oxide TFT is commonly used because of the negligibly small parasitic capacitance. The drain current of the dual-gate TFT operated in the bulk accumulation mode TFT is three to six times higher than that of the single-gate TFT, which leads to the low operating voltage and low power consumption. However, the fabrication process of the dual-gate TFT is more complicated than the single-gate TFT.

The main advantage of the solution-processed oxide TFT is the low manufacturing cost. Spin coating has been widely studied but the uniformity is a concern. Spray pyrolysis can provide good uniformity for OS. The as-deposited metal-oxide film can be amorphous or polycrystalline depending on the substrate temperature. The polycrystalline oxide TFT can have a higher mobility than the amorphous oxide TFT because of its higher mass density and fewer oxygen vacancies and OH groups. Inkjet printing has an advantage of selective deposition of oxide semiconductor on the channel region. Currently, the performance of the inkjet oxide TFT is not as good as that of the sputtered oxide TFT.

It is concluded that for the future generation displays, there is no single standard process that can provide high performance oxide TFTs. It is generally recognized that for low power consumption, the channel length should be short and the high-k GI needs to be used. Currently, no optimized high-k GI material or process condition has been identified. Therefore, it is necessary to establish a standard oxide TFT fabrication process for production. Note that for displays, the process steps of the oxide TFT backplane can be similar to those of the a-Si:H, but simpler than those of the LTPS backplane. The oxide TFT fabrication process varies with the display company. This chapter reviewed and discussed the effects of the fabrication process and the device structure of oxide TFTs that are used currently or can be used in the future in AMOLED and AM micro-LED displays.

References

1 Yabuta, H., Sano, M., Abe, K. et al. (2006). High-mobility thin-film transistor with amorphous $InGaZnO_4$ channel fabricated by room temperature rf-magnetron sputtering. *Applied Physics Letters* 89 (11): 112123.

2 Kim, W., Lee, S.-H., Bang, J.-H. et al. (2011). Role of O_2/Ar mixing ratio on the performances of IZO thin film transistors fabricated using a two-step deposition process. *Thin Solid Films* 520 (5): 1475–1478.

3 Glushkova, A.V., Dekkers, H.F.W., Nag, M. et al. (2021). Systematic study on the amorphous, C-axis-aligned crystalline, and protocrystalline phases in In–Ga–Zn oxide thin-film transistors. *ACS Applied Electronic Materials* 3 (3): 1268–1278.

4 Paine, D.C., Yaglioglu, B., Beiley, Z. et al. (2008). Amorphous IZO-based transparent thin film transistors. *Thin Solid Films* 516 (17): 5894–5898.

5 Chowdhury, M.D.H., Migliorato, P., Jang, J. et al. (2010). Light induced instabilities in amorphous indium–gallium–zinc–oxide thin-film transistors. *Applied Physics Letters* 97 (17): 173506.

6 Chen, X.F., He, G., Gao, J. et al. (2015). Substrate temperature dependent structural, optical and electrical properties of amorphous InGaZnO thin films. *Journal of Alloys and Compounds* 632: 533–539.

7 Lee, J., Choi, B. et al. (2018). Effects of channel type and doping on hysteresis in low-temperature poly-si thin-film transistors. *IEEE Transactions on Electron Devices* 65 (3): 986–994.

8 Jung, C.H., Kang, H.I., Yoon, D.H. et al. (2013). The electrical, optical, and structural properties of amorphous indium gallium zinc oxide films and channel thin-film transistors. *Solid-State Electronics* 79: 125–129.

9 Wang, H., He, J., Xu, Y. et al. (2020). Impact of hydrogen dopant incorporation on InGaZnO, ZnO and In_2O_3 thin film transistors. *Physical Chemistry Chemical Physics* 22 (3): 1591–1597.

10 Li, G., Abliz, A., Xu, L. et al. (2018). Understanding hydrogen and nitrogen doping on active defects in amorphous In-Ga-Zn-O thin film transistors. *Applied Physics Letters* 112 (25): 253504.

11 Miyase, T., Watanabe, K., Sakaguchi, I. et al. (2014). Roles of hydrogen in amorphous oxide semiconductor In-Ga-Zn-O: comparison of conventional and ultra-high-vacuum sputtering. *ECS Journal of Solid State Science and Technology* 3 (9): Q3085.

12 Pan, W., Wang, Y., Wang, Y. et al. (2023). Multiple effects of hydrogen on InGaZnO thin-film transistor and the hydrogenation-resistibility enhancement. *Journal of Alloys and Compounds* 947: 169509.

13 Velichko, R., Magari, Y., Furuta, M. et al. (2022). Defect passivation and carrier reduction mechanisms in hydrogen-doped In-Ga-Zn-O (IGZO:H) films upon low-temperature annealing for flexible device applications. *Materials* 15 (1): 334.

14 Lee, S., Choi, M., Jang, J. et al. (2023). Highly stable, self-aligned coplanar a-InGaZnO TFTs with oxide and hydrogenated amorphous Si contacts. *IEEE Electron Device Letters* 44 (12): 1983–1986.

15 Park, J., Kim, S., Kim, C. et al. (2008). High-performance amorphous gallium indium zinc oxide thin-film transistors through N_2O plasma passivation. *Applied Physics Letters* 93 (5): 053505.

16 Hsieh, T.-Y., Chang, T.-C., Chen, T.-C. et al. (2011). Effect of N_2O plasma treatment on the improvement of instability under light illumination for InGaZnO thin-film transistors. *Thin Solid Films* 520 (5): 1427–1431.

17 Jhu, J.-C., Chang, T.-C., Chang, G.-W. et al. (2013). Reduction of defect formation in amorphous indium-gallium-zinc-oxide thin film transistors by N_2O plasma treatment. *Journal of Applied Physics* 114 (20): 204501.

18 Zhou, X., Zhang, X., Shao, Y. et al. (2018). P-6.1: asymmetric effects of gate-bias stress voltage on the stability under positive and negative gate-bias stress of a-IGZO TFTs. *SID Symposium Digest of Technical Papers* 49: 597–600.

19 Kim, H., Han, C., Kim, D. et al. (2023). Electrical performance and reliability enhancement of a-IGZO TFTs via post-N_2O plasma optimization. *IEEE Transactions on Electron Devices* 70 (7): 3611–3616.

20 Tsai, C.-T., Chang, T.-C., Chen, S.-C. et al. (2010). Influence of positive bias stress on N_2O plasma improved InGaZnO thin film transistor. *Applied Physics Letters* 96 (24): 242105.

21 Sheng, J., Park, J., Choi, D. et al. (2016). A study on the electrical properties of atomic layer deposition grown InO_x on flexible substrates with respect to N_2O plasma treatment and the associated thin-film transistor behavior under repetitive mechanical stress. *ACS Applied Materials & Interfaces* 8 (45): 31136–31143.

22 Li, J., Zhang, Y., Wang, J. et al. (2023). Near-ideal top-gate controllability of InGaZnO thin-film transistors by suppressing interface defects with an ultrathin atomic layer deposited gate insulator. *ACS Applied Materials & Interfaces* 15 (6): 8666–8675.

23 Rabbi, M.H., Lee, S., Sasaki, D. et al. (2022). Polycrystalline InGaO thin-film transistors with coplanar structure exhibiting average mobility of $\approx 78\,cm^2\,V^{-1}\,s^{-1}$ and excellent stability for replacing current poly-Si thin-film transistors for organic light-emitting diode displays. *Small Methods* 6 (9): 2200668.

24 Rabbi, M.H., Billah, M.M., Siddik, A.B. et al. (2020). Extremely stable dual gate coplanar amorphous InGaZnO thin film transistor with split active layer by N_2O annealing. *IEEE Electron Device Letters* 41 (12): 1782–1785.

25 Liang, T., Zhang, X., Zhou, X. et al. (2017). P-20: effects of N_2O plasma treatment time on the performance of self-aligned top-gate amorphous oxide thin film transistors. *SID Symposium Digest of Technical Papers* 48: 1299–1302.

26 Yim, H., Kim, D.-H., Choi, S. et al. (2011). P-20: highly stable amorphous indium gallium zinc oxide thin-film transistors with N_2O plasma treatment. *SID Symposium Digest of Technical Papers* 42 (1): 1170–1172.

27 Park, J.C., Kim, S.W., Kim, S.I. et al. (2009). High performance amorphous oxide thin film transistors with self-aligned top-gate structure. *2009 IEEE International Electron Devices Meeting (IEDM)*, Baltimore, MD, USA (07–09 December 2009).

28 Park, J.-S., Jeong, J.K., and Mo, Y.-G. (2007). Improvements in the device characteristics of amorphous indium gallium zinc oxide thin-film transistors by Ar plasma treatment. *Applied Physics Letters* 90 (26): 262106.

29 Rahaman, A., Billah, M.M., Um, J.G. et al. (2018). Effect of doping fluorine in offset region on performance of coplanar a-IGZO TFTs. *IEEE Electron Device Letters* 39 (9): 1318–1321.

30 Um, J.G., Jang, J. et al. (2018). Heavily doped n-type a-IGZO by F plasma treatment and its thermal stability up to 600 °C. *Applied Physics Letters* 112 (16): 162104.

31 Kim, M.-M., Kim, M.-H., and Ryu, S.-M. (2015). Coplanar homojunction a-InGaZnO thin film transistor fabricated using ultraviolet irradiation. *RSC Advances* 5 (101): 82947–82951.

32 Kim, M.-H., Choi, S.-Y., Jeon, S.-H. et al. (2018). Stability behavior of self-aligned coplanar a-IGZO thin film transistors fabricated by deep ultraviolet irradiation. *ECS Journal of Solid State Science and Technology* 7 (4): Q60.

33 Magari, Y., Makino, H., Furuta, M. et al. (2017). Carrier generation mechanism and origin of subgap states in Ar- and He-plasma-treated In–Ga–Zn–O thin films. *ECS Journal of Solid State Science and Technology* 6 (8): Q101.

34 Jeong, H., Lee, B., Lee, Y. et al. (2014). Coplanar amorphous-indium-gallium-zinc-oxide thin film transistor with He plasma treated heavily doped layer. *Applied Physics Letters* 104 (2): 022115.

35 Ahn, B.D., Shin, H.S., Kim, H.J. et al. (2008). Comparison of the effects of Ar and H_2 plasmas on the performance of homojunctioned amorphous indium gallium zinc oxide thin film transistors. *Applied Physics Letters* 93 (20): 203506.

36 Park, H., Yun, J., Park, S. et al. (2022). Enhancing the contact between a-IGZO and metal by hydrogen plasma treatment for a high-speed varactor (>30 GHz). *ACS Applied Electronic Materials* 4 (4): 1769–1775.

37 Sato, A., Shimada, M., Abe, K. et al. (2009). Amorphous In–Ga–Zn-O thin-film transistor with coplanar homojunction structure. *Thin Solid Films* 518 (4): 1309–1313.

38 Yang, H., Zhou, X., Fu, H. et al. (2021). Metal reaction-induced bulk-doping effect in forming conductive source-drain regions of self-aligned top-gate amorphous InGaZnO thin-film transistors. *ACS Applied Materials & Interfaces* 13 (9): 11442–11448.

39 Peng, H., Chang, B., Fu, H. et al. (2020). Top-gate amorphous indium-gallium-zinc-oxide thin-film transistors with magnesium metallized source/drain regions. *IEEE Transactions on Electron Devices* 67 (4): 1619–1624.

40 Abliz, A., Nurmamat, P., Wan, D. et al. (2023). Rational design of oxide heterostructure InGaZnO/TiO_2 for high-performance thin-film transistors. *Applied Surface Science* 609: 155257.

41 Mativenga, M., Haque, F., Billah, M.M. et al. (2021). Origin of light instability in amorphous IGZO thin-film transistors and its suppression. *Scientific Reports* 11 (1): 14618.

42 Um, J.G., Mativenga, M., Migliorato, P. et al. (2012). Increase of interface and bulk density of states in amorphous-indium-gallium-zinc-oxide thin-film transistors with negative-bias-under-illumination-stress time. *Applied Physics Letters* 101 (11): 113504.

43 Jeon, J.K., Um, J.G., Lee, S. et al. (2017). Control of O-H bonds at a-IGZO/SiO_2 interface by long time thermal annealing for highly stable oxide TFT. *AIP Advances* 7 (12): 125110.

44 Aman, S.G.M., Koretomo, D., Magari, Y. et al. (2018). Influence of deposition temperature and source gas in PE-CVD for SiO_2 passivation on performance and reliability of In–Ga–Zn–O thin-film transistors. *IEEE Transactions on Electron Devices* 65 (8): 3257–3263.

45 Bae, J., Ali, A., Islam, M.M. et al. (2023). As-grown crystalline InGaZnO by spray pyrolysis on a flexible substrate for a thin-film transistor with excellent stability. *ACS Applied Materials & Interfaces* 15 (33): 39494–39504.

46 Jeong, H., Nam, S., Park, K. et al. (2020). Significant performance and stability improvements of low-temperature IGZO TFTs by the formation of In-F nanoparticles on an SiO_2 buffer layer. *Nanomaterials* 10 (6): 1165.

47 Zarchi, M., Ahangarani, S., Sanjari, M.Z. et al. (2014). Properties of silicon dioxide film deposited by PECVD at low temperature/pressure. *Metallurgical and Materials Engineering* 20: 89–96.

48 Kattamis, A., Cheng, I.-C., Allen, S. et al. (2004). Hydrogen in ultralow temperature SiO_2 for nanocrystalline silicon thin film transistors. *MRS Online Proceedings Library* 814 (1): 30–35.

49 Herth, E., Legrand, B., Buchaillot, L. et al. (2010). Optimization of SiN_X:H films deposited by PECVD for reliability of electronic, microsystems, and optical applications. *Microelectronics Reliability* 50 (8): 1103–1106.

50 Wu, C.-H., Hsieh, H.-H., Chien, C.-W. et al. (2009). Self-aligned top-gate coplanar In-Ga-Zn-O thin-film transistors. *Journal of Display Technology* 5 (12): 515–519.

51 Hong, S., Lee, S., Mativenga, M. et al. (2014). Reduction of negative bias and light instability of a-IGZO TFTs by dual-gate driving. *IEEE Electron Device Letters* 35 (1): 93–95.

52 Chun, M., Chowdhury, M.D.H., Jang, J. et al. (2015). Semiconductor to metallic transition in bulk accumulated amorphous indium-gallium-zinc-oxide dual gate thin-film transistor. *AIP Advances* 5 (5): 057165.

53 Chowdhury, M.D.H., Mativenga, M., Um, J.G. et al. (2015). Effect of SiO_2 and SiO_2/SiN_x passivation on the stability of amorphous indium-gallium zinc-oxide thin-film transistors under high humidity. *IEEE Transactions on Electron Devices* 62 (3): 869–874.

54 Liu, S.-E., Yu, M.-J., Lin, C.-Y. et al. (2011). Influence of passivation layers on characteristics of a-InGaZnO thin-film transistors. *IEEE Electron Device Letters* 32 (2): 161–163.

55 Siddik, A.B., Billah, M.M., Lee, S. et al. (2021). Blue laser annealed sub-micron channel P-type low temperature poly-Si TFT without kink effect for high-resolution display. *IEEE Electron Device Letters* 42 (2): 172–175.

56 Chun, M., Um, J.G., Park, M.S. et al. (2016). Effect of top gate potential on bias-stress for dual gate amorphous indium-gallium-zinc-oxide thin film transistor. *AIP Advances* 6 (7): 075217.

57 Jeong, S.-G., Jeong, H.-J., Choi, W.-H. et al. (2020). Hydrogen impacts of PEALD InGaZnO TFTs using SiO_x gate insulators deposited by PECVD and PEALD. *IEEE Transactions on Electron Devices* 67 (10): 4250–4255.

58 Park, J., Kim, H., Choi, P. et al. (2021). Effect of ALD- and PEALD-grown Al_2O_3 gate insulators on electrical and stability properties for a-IGZO thin-film transistor. *Electronic Materials Letters* 17 (4): 299–306.

59 Castillo-Saenz, J., Nedev, N., Valdez-Salas, B. et al. (2021). Properties of Al_2O_3 thin films grown by PE-ALD at low temperature using H_2O and O_2 plasma oxidants. *Coatings* 11 (10): 1266.

60 Shao, Y., Wu, X., Zhang, M.-N. et al. (2019). High-performance a-InGaZnO thin-film transistors with extremely low thermal budget by using a hydrogen-rich Al_2O_3 dielectric. *Nanoscale Research Letters* 14 (1): 122.

61 Mativenga, M., Choi, J.W., Hur, J.H. et al. (2011). Highly stable amorphous indium–gallium–zinc-oxide thin-film transistor using an etch-stopper and a via-hole structure. *Journal of Information Display* 12 (1): 47–50.

62 Kim, M., Jeong, J.H., Lee, H.J. et al. (2007). High mobility bottom gate InGaZnO thin film transistors with SiO_x etch stopper. *Applied Physics Letters* 90 (21): 212114.

63 Toda, T., Wang, D., Jiang, J. et al. (2014). Quantitative analysis of the effect of hydrogen diffusion from silicon oxide etch-stopper layer into amorphous In–Ga–Zn–O on thin-film transistor. *IEEE Transactions on Electron Devices* 61 (11): 3762–3767.

64 Xia, Z., Lu, L., Li, J. et al. (2018). A bottom-gate metal–oxide thin-film transistor with self-aligned source/drain regions. *IEEE Transactions on Electron Devices* 65 (7): 2820–2826.

65 Mativenga, M., Geng, D., Chang, J.H. et al. (2012). Performance of 5-nm a-IGZO TFTs with various channel lengths and an etch stopper manufactured by back UV exposure. *IEEE Electron Device Letters* 33 (6): 824–826.

66 Geng, D., Kang, D.H., Jang, J. et al. (2011). High-performance amorphous indium–gallium–zinc–oxide thin-film transistor with a self-aligned etch stopper patterned by back-side UV exposure. *IEEE Electron Device Letters* 32 (6): 758–760.

67 Park, S.-H.K., Ryu, M.-K., Oh, H. et al. (2013). Double-layered passivation film structure of Al_2O_3/SiN_x for high mobility oxide thin film transistors. *Journal of Vacuum Science & Technology B* 31 (2): 020601.

68 Oh, S., Baeck, J.H., Shin, H.S. et al. (2014). Comparison of top-gate and bottom-gate amorphous InGaZnO thin-film transistors with the same SiO_2/a-InGaZnO/SiO_2 stack. *IEEE Electron Device Letters* 35 (10): 1037–1039.

69 Kang, D.H., Kang, I., Ryu, S.H. et al. (2011). Self-aligned coplanar a-IGZO TFTs and application to high-speed circuits. *IEEE Electron Device Letters* 32 (10): 1385–1387.

70 Geng, D., Kang, D.H., Seok, M.J. et al. (2012). High-speed and low-voltage-driven shift register with self-aligned coplanar a-IGZO TFTs. *IEEE Electron Device Letters* 33 (7): 1012–1014.

71 Li, X., Xin, E., Chen, L. et al. (2013). Effect of etching stop layer on characteristics of amorphous IGZO thin film transistor fabricated at low temperature. *AIP Advances* 3 (3): 032137.

72 Koretomo, D., Toda, T., Matsuda, T. et al. (2016). Anomalous increase in field-effect mobility in In–Ga–Zn–O thin-film transistors caused by dry-etching damage through etch-stop layer. *IEEE Transactions on Electron Devices* 63 (7): 2785–2789.

73 You, B.H., Lee, B.J., Lee, K.-C. et al. (2008). 55.1: 12.1-inch a-Si:H TFT LCD with embedded touch screen panel. *SID Symposium Digest of Technical Papers* 39 (1): 830–833.

74 Nag, M., Bhoolokam, A., Steudel, S. et al. (2014). Back-channel-etch amorphous indium–gallium–zinc oxide thin-film transistors: the impact of source/drain metal etch and final passivation. *Japanese Journal of Applied Physics* 53 (11): 111401.

75 Ryu, S.H., Park, Y.C., Mativenga, M. et al. (2012). Amorphous-$InGaZnO_4$ thin-film transistors with damage-free back channel wet-etch process. *ECS Solid State Letters* 1 (2): Q17.

76 Park, Y.C., Um, J.G., Mativenga, M. et al. (2015). Modification of electrode-etchant for sidewall profile control and reduced back-channel corrosion of inverted-staggered metal-oxide TFTs. *ECS Journal of Solid State Science and Technology* 4 (12): Q124.

77 Choi, S.-H., Han, M.-K. et al. (2012). Effect of deposition temperature of SiOx passivation layer on the electrical performance of a-IGZO TFTs. *IEEE Electron Device Letters* 33 (3): 396–398.

78 Park, J.C., Ahn, S.-E., Lee, H.-N. et al. (2013). High-performance low-cost back-channel-etch amorphous gallium–indium–zinc oxide thin-film transistors by curing and passivation of the damaged back channel. *ACS Applied Materials & Interfaces* 5 (23): 12262–12267.

79 Liu, X., Wang, L.L., Hu, H. et al. (2015). Performance and stability improvements of back-channel-etched amorphous indium–gallium–zinc thin-film-transistors by CF_4+O_2 plasma treatment. *IEEE Electron Device Letters* 36 (9): 911–913.

80 Ji, K.H., Kim, J.-I., Mo, Y.-G. et al. (2010). Comparative study on light-induced bias stress instability of IGZO transistors with SiN_x and SiO_2 gate dielectrics. *IEEE Electron Device Letters* 31 (12): 1404–1406.

81 Ahn, B.D., Shin, H.S., Kim, G.H. et al. (2009). A novel amorphous InGaZnO thin film transistor structure without source/drain layer deposition. *Japanese Journal of Applied Physics* 48 (3S2): 03B019.

82 Baek, G., Abe, K., Kumomi, H. et al. (2013). Scaling of coplanar homojunction amorphous In–Ga–Zn–O thin-film transistors. *Japanese Journal of Applied Physics* 52 (3S): 03BB05.

83 Kim, H.W., Kim, E.S., Park, J.S. et al. (2018). Influence of effective channel length in self-aligned coplanar amorphous-indium-gallium-zinc-oxide thin-film transistors with different annealing temperatures. *Applied Physics Letters* 113 (2): 022104.

84 Kang, D.H., Han, J.U., Mativenga, M. et al. (2013). Threshold voltage dependence on channel length in amorphous-indium-gallium-zinc-oxide thin-film transistors. *Applied Physics Letters* 102 (8): 083508.

85 Um, J.K., Lee, S., Jin, S. et al. (2015). High-performance homojunction a-IGZO TFTs with selectively defined low-resistive a-IGZO source/drain electrodes. *IEEE Transactions on Electron Devices* 62 (7): 2212–2218.

86 Park, J., Song, I., Kim, S. et al. (2008). Self-aligned top-gate amorphous gallium indium zinc oxide thin film transistors. *Applied Physics Letters* 93 (5): 053501.

87 Ha, S.H., Kang, D.H., Kang, I. et al. (2013). Channel length dependent bias-stability of self-aligned coplanar a-IGZO TFTs. *Journal of Display Technology* 9 (12): 985–988.

88 Ha, C., Lee, H.-J., Kwon, J.-W. et al. (2015). 69.2: Distinguished paper: high reliable a-IGZO TFTs with self-aligned coplanar structure for large-sized ultrahigh-definition OLED TV. *SID Symposium Digest of Technical Papers* 46 (1): 1020–1022.

89 Shin, H.-J., Choi, S.-H., Kim, D.-M. et al. (2021). 45-1: A novel 88-inch 8K OLED display for premium large-size TVs. *SID Symposium Digest of Technical Papers* 52 (1): 611–614.

90 Kang, D.H., Kang, I., Ryu, S.H. et al. (2013). Effect of SiO_2 and/or SiN_x passivation layer on thermal stability of self-aligned coplanar amorphous indium–gallium–zinc–oxide thin-film transistors. *Journal of Display Technology* 9 (9): 699–703.

91 Park, S.H., Mativenga, M., Jang, J. et al. (2013). Increase of mobility in dual gate amorphous-$InGaZnO_4$ thin-film transistors by pseudo-doping. *Applied Physics Letters* 103 (4): 043509.

92 Li, X., Geng, D., Mativenga, M. et al. (2014). High-speed dual-gate a-IGZO TFT-based circuits with top-gate offset structure. *IEEE Electron Device Letters* 35 (4): 461–463.

93 Mativenga, M., An, S., Jang, J. et al. (2013). Bulk accumulation a-IGZO TFT for high current and turn-on voltage uniformity. *IEEE Electron Device Letters* 34 (12): 1533–1535.

94 Son, K.-S., Jung, J.-S., Lee, K.-H. et al. (2010). Characteristics of double-gate Ga–In–Zn–O thin-film transistor. *IEEE Electron Device Letters* 31 (3): 219–221.

95 Chen, Y., Geng, D., Mativenga, M. et al. (2015). High-speed pseudo-CMOS circuits using bulk accumulation a-IGZO TFTs. *IEEE Electron Device Letters* 36 (2): 153–155.

96 Lee, S., Mativenga, M., Jang, J. et al. (2014). Removal of negative-bias-illumination-stress instability in amorphous-InGaZnO thin-film transistors by top-gate offset structure. *IEEE Electron Device Letters* 35 (9): 930–932.

97 Chen, Y., Geng, D., Lin, T. et al. (2016). Full-swing clock generating circuits on plastic using a-IGZO dual-gate TFTs with pseudo-CMOS and bootstrapping. *IEEE Electron Device Letters* 37 (7): 882–885.

98 Seok, M.J., Mativenga, M., Geng, D. et al. (2013). Achieving high performance oxide TFT-based inverters by use of dual-gate configurations with floating and biased secondary gates. *IEEE Transactions on Electron Devices* 60 (11): 3787–3793.

99 Baek, G., Abe, K., Kuo, A. et al. (2011). Electrical properties and stability of dual-gate coplanar homojunction DC sputtered amorphous indium–gallium–zinc–oxide thin-film transistors and its application to AM-OLEDs. *IEEE Transactions on Electron Devices* 58 (12): 4344–4353.

100 Son, K.-S., Jung, J.-S., Lee, K.-H. et al. (2010). Highly stable double-gate Ga–In–Zn–O thin-film transistor. *IEEE Electron Device Letters* 31 (8): 812–814.

101 MacDonald, B.A., Rollins, K., MacKerron, D. et al. (2005). Engineered films for display technologies. In: *Flexible Flat Panel Displays* (ed. G.P. Crawford), 11–33. John Wiley & Sons Ltd.

102 Gregg, A., York, L., Strnad, M. et al. (2005). Roll-to-roll manufacturing of flexible displays. In: *Flexible Flat Panel Displays* (ed. G.P. Crawford), 409–445. John Wiley & Sons Ltd.

103 Allen, K.J. (2005). Reel to real: prospects for flexible displays. *Proceedings of the IEEE* 93 (8): 1394–1399.

104 Serikawa, T. and Omata, F. (2002). High-quality polycrystalline Si TFTs fabricated on stainless-steel foils by using sputtered Si films. *IEEE Transactions on Electron Devices* 49 (5): 820–825.

105 Jin, D.-U., Lee, J.-S., Kim, T.-W. et al. (2009). 65.2: Distinguished paper: world-largest (6.5″) flexible full color top emission AMOLED display on plastic film and its bending properties. *SID Symposium Digest of Technical Papers* 40 (1): 983–985.

106 Theiss, S.D., Wagner, S. et al. (1996). Amorphous silicon thin-film transistors on steel foil substrates. *IEEE Electron Device Letters* 17 (12): 578–580.

107 Kim, S.Y., Kim, K., Hong, K. et al. (2009). P-107: Mechanism of peel-off of metal substrate for flexible devices. *SID Symposium Digest of Technical Papers* 40 (1): 1516–1519.

108 French, I. et al. (2009). 10.2: Invited paper: flexible E-books. *SID Symposium Digest of Technical Papers* 40 (1): 100–103.

109 Lee, C.-C., Chang, Y.-Y., Cheng, H.-C. et al. (2010). 54.1: Invited paper: a novel approach to make flexible active matrix displays. *SID Symposium Digest of Technical Papers* 41 (1): 810–813.

110 Jo, J.-W., Kim, J., Kim, K.-T. et al. (2015). Highly stable and imperceptible electronics utilizing photoactivated heterogeneous sol-gel metal–oxide dielectrics and semiconductors. *Advanced Materials* 27 (7): 1182–1188.

111 Mativenga, M., Choi, M.H., Choi, J.W. et al. (2011). Transparent flexible circuits based on amorphous-indium–gallium–zinc–oxide thin-film transistors. *IEEE Electron Device Letters* 32 (2): 170–172.

112 Lee, S., Shin, J., Jang, J. et al. (2017). Top interface engineering of flexible oxide thin-film transistors by splitting active layer. *Advanced Functional Materials* 27 (11): 1604921.

113 Lee, S., Chen, Y., Kim, J. et al. (2018). Transparent AMOLED display driven by split oxide TFT backplane. *Journal of the Society for Information Display* 26 (3): 164–168.

114 Lee, S., Jeong, D., Mativenga, M. et al. (2017). Highly robust bendable oxide thin-film transistors on polyimide substrates via mesh and strip patterning of device layers. *Advanced Functional Materials* 27 (29): 1700437.

115 Kim, Y.-H., Lee, E., Um, J.G. et al. (2016). Highly robust neutral plane oxide TFTs withstanding 0.25 mm bending radius for stretchable electronics. *Scientific Reports* 6 (1): 25734.

116 Nomura, K., Ohta, H., Takagi, A. et al. (2004). Room-temperature fabrication of transparent flexible thin-film transistors using amorphous oxide semiconductors. *Nature* 432: 488–492.

117 Fortunato, E., Barquinha, P., Martins, R. et al. (2012). Oxide semiconductor thin-film transistors: a review of recent advances. *Advanced Materials* 24 (22): 2945–2986.

118 Xu, B., Akhtar, A., Liu, Y. et al. (2016). An epidermal stimulation and sensing platform for sensorimotor prosthetic control, management of lower back exertion, and electrical muscle activation. *Advanced Materials* 28 (22): 4462–4471.

119 Huang, X., Liu, Y., Chen, K. et al. (2014). Stretchable, wireless sensors and functional substrates for epidermal characterization of sweat. *Small* 10 (15): 3083–3090.

120 Jayathilaka, W.A.D.M., Qi, K., Qin, Y. et al. (2019). Significance of nanomaterials in wearables: a review on wearable actuators and sensors. *Advanced Materials* 31 (7): 1805921.

121 Park, S., Cho, K., Oh, H. et al. (2016). Electrical and mechanical characteristics of fully transparent IZO thin-film transistors on stress-relieving bendable substrates. *Applied Physics Letters* 109 (14): 143504.

122 Lee, S.J., Ko, J., Nam, K.-H. et al. (2016). Fully solution-processed and foldable metal-oxide thin-film transistor. *ACS Applied Materials & Interfaces* 8 (20): 12894–12900.

123 Eda, G., Mattevi, C., Yamaguchi, H. et al. (2009). Insulator to semimetal transition in graphene oxide. *The Journal of Physical Chemistry C* 113 (35): 15768–15771.

124 Sun, D.-M., Liu, C., Ren, W.-C. et al. (2016). All-carbon thin-film transistors as a step towards flexible and transparent electronics. *Advanced Electronic Materials* 2 (11): 1600229.

125 Park, S.-I., Le, A.-P., Wu, J. et al. (2010). Light emission characteristics and mechanics of foldable inorganic light-emitting diodes. *Advanced Materials* 22 (28): 3062–3066.

126 Um, J.G., Geng, D., Mativenga, M. et al. (2016). 64-3: Distinguished student paper: bulk accumulation oxide TFTs for flexible AMOLED display with high yield integrated gate driver. *SID Symposium Digest of Technical Papers* 47 (1): 872–875.

127 Geng, D., Chen, Y.F., Mativenga, M. et al. (2015). 30 μm-Pitch oxide TFT-based gate driver design for small-size, high-resolution, and narrow-bezel displays. *IEEE Electron Device Letters* 36 (8): 805–807.

128 Billah, M.M., Hasan, M.M., Jang, J. et al. (2017). Effect of tensile and compressive bending stress on electrical performance of flexible a-IGZO TFTs. *IEEE Electron Device Letters* 38 (7): 890–893.

129 Gleskova, H., Wagner, S., Gašparík, V. et al. (2001). 150°C Amorphous silicon thin-film transistor technology for polyimide substrates. *Journal of the Electrochemical Society* 148 (7): G370.

130 Hosono, H., Yasukawa, M., Kawazoe, H. et al. (1996). Novel oxide amorphous semiconductors: transparent conducting amorphous oxides. *Journal of Non-Crystalline Solids* 203: 334–344.

131 Lin, C.-L., Chang, W.-Y., Hung, C.-C. et al. (2013). Compensating pixel circuit driving AMOLED display with a-IGZO TFTs. *IEEE Electron Device Letters* 34 (9): 1166–1168.

132 Park, J.S., Maeng, W.-J., Kim, H.-S. et al. (2012). Review of recent developments in amorphous oxide semiconductor thin-film transistor devices. *Thin Solid Films* 520 (6): 1679–1693.

133 Cherenack, K.H., Munzenrieder, N.S., Troster, G. et al. (2010). Impact of mechanical bending on ZnO and IGZO thin-film transistors. *IEEE Electron Device Letters* 31 (11): 1254–1256.

134 Inoue, S., Utsunomiya, S., Saeki, T. et al. (2002). Surface-free technology by laser annealing (SUFTLA) and its application to poly-Si TFT-LCDs on plastic film with integrated drivers. *IEEE Transactions on Electron Devices* 49 (8): 1353–1360.

135 Li, X., Ono, T., Wang, Y. et al. (2003). Ultrathin single-crystalline-silicon cantilever resonators: fabrication technology and significant specimen size effect on Young's modulus. *Applied Physics Letters* 83 (15): 3081–3083.

136 Wang, J.-S., Feng, X.-Q., Wang, G.-F. et al. (2008). Twisting of nanowires induced by anisotropic surface stresses. *Applied Physics Letters* 92 (19): 191901.

137 Wu, B., Heidelberg, A., Boland, J.J. et al. (2005). Mechanical properties of ultrahigh-strength gold nanowires. *Nature Materials* 4 (7): 525–529.

138 Lucas, M., Mai, W., Yang, R. et al. (2007). Aspect ratio dependence of the elastic properties of ZnO nanobelts. *Nano Letters* 7 (5): 1314–1317.

139 Zheng, X.-P., Cao, Y.-P., Li, B. et al. (2010). Surface effects in various bending-based test methods for measuring the elastic property of nanowires. *Nanotechnology* 21 (20): 205702.

140 Mativenga, M., Hong, S., Jang, J. et al. (2013). High current stress effects in amorphous-$InGaZnO_4$ thin-film transistors. *Applied Physics Letters* 102 (2): 023503.

141 Munzenrieder, N., Cherenack, K.H., Troster, G. et al. (2011). The effects of mechanical bending and illumination on the performance of flexible IGZO TFTs. *IEEE Transactions on Electron Devices* 58 (7): 2041–2048.

142 Kohan, A.F., Ceder, G., Morgan, D. et al. (2000). First-principles study of native point defects in ZnO. *Physical Review B* 61 (22): 15019–15027.

143 Tai, Y.-H., Chiu, H.-L., Chou, L.-S. et al. (2012). The deterioration of a-IGZO TFTs owing to the copper diffusion after the process of the source/drain metal formation. *Journal of the Electrochemical Society* 159 (5): J200.

144 Sharma, B.K., Jang, B., Lee, J.E. et al. (2013). Load-controlled roll transfer of oxide transistors for stretchable electronics. *Advanced Functional Materials* 23 (16): 2024–2032.

145 Heremans, P., Tripathi, A.K., de, de Meux, A., J. et al. (2016). Mechanical and electronic properties of thin-film transistors on plastic, and their integration in flexible electronic applications. *Advanced Materials* 28 (22): 4266–4282.

146 Liao, P.-Y., Chang, T.-C., Su, W.-C. et al. (2016). Effect of mechanical-strain-induced defect generation on the performance of flexible amorphous In–Ga–Zn–O thin-film transistors. *Applied Physics Express* 9 (12): 124101.

147 Migliorato, P., Delwar, M.D.H., Um, J.G. et al. (2012). Light/negative bias stress instabilities in indium gallium zinc oxide thin film transistors explained by creation of a double donor. *Applied Physics Letters* 101 (12): 123502.

148 Kim, H.-J., Kim, Y.-J. et al. (2014). Influence of external forces on the mechanical characteristics of the a-IGZO and graphene based flexible display. *IOP Conference Series: Materials Science and Engineering* 62 (1): 012022.

149 Yang, J., Gaspar, J., Paul, O. et al. (2008). Fracture properties of LPCVD silicon nitride and thermally grown silicon oxide thin films from the load-deflection of long Si_3N_4 and SiO_2/Si_3N_4 diaphragms. *Journal of Microelectromechanical Systems* 17 (5): 1120–1134.

150 Tsukamoto, H. (2014). Mechanical properties of zirconia-titanium composites. *International Journal of Materials Science and Applications* 3 (5): 260–267.

151 Neerinck, D.G., Vink, T.J. et al. (1996). Depth profiling of thin ITO films by grazing incidence X-ray diffraction. *Thin Solid Films* 278 (1): 12–17.

152 Jeong, J.K., Jeong, J.H., Yang, H.W. et al. (2007). High performance thin film transistors with cosputtered amorphous indium gallium zinc oxide channel. *Applied Physics Letters* 91 (11): 113505.

153 Yim, J.-R., Jung, S.-Y., Yeon, H.-W. et al. (2012). Effects of metal electrode on the electrical performance of amorphous In–Ga–Zn–O thin film transistor. *Japanese Journal of Applied Physics* 51 (1R): 011401.

154 Na, J.H., Kitamura, M., Arakawa, Y. et al. (2008). High field-effect mobility amorphous InGaZnO transistors with aluminum electrodes. *Applied Physics Letters* 93 (6): 063501.

155 Kim, W.-S., Moon, Y.-K., Kim, K.-T. et al. (2010). An investigation of contact resistance between metal electrodes and amorphous gallium–indium–zinc oxide (a-GIZO) thin-film transistors. *Thin Solid Films* 518 (22): 6357–6360.

156 Han, D.-S., Kang, Y.-J., Park, J.-H. et al. (2014). Influence of molybdenum source/drain electrode contact resistance in amorphous zinc–tin-oxide (a-ZTO) thin film transistors. *Materials Research Bulletin* 58: 174–177.

157 Yun, P.S., Koike, J. et al. (2011). Metal reaction doping and ohmic contact with Cu-Mn electrode on amorphous In-Ga-Zn-O semiconductor. *Journal of the Electrochemical Society* 158 (10): H1034.

158 Kim, H.S., Byrne, P.D., Facchetti, A. et al. (2008). High performance solution-processed indium oxide thin-film transistors. *Journal of the American Chemical Society* 130 (38): 12580–12581.

159 Lee, D.H., Park, S.M., Yang, J.I. et al. (2013). Effect of chloride precursors on the stability of solution-processed indium zinc oxide thin-film transistors. *Japanese Journal of Applied Physics* 52 (10S): 10MA02.

160 Kim, Y.G., Kim, T., Avis, C. et al. (2016). Stable and high-performance indium oxide thin-film transistor by Ga doping. *IEEE Transactions on Electron Devices* 63 (3): 1078–1084.

161 Rabbi, M.H., Ali, A., Park, C. et al. (2023). High performance amorphous $In_{0.5}Ga_{0.5}O$ thin-film transistor embedded with nanocrystalline In_2O_3 dots for flexible display application. *Advanced Electronic Materials* 9 (10): 2300169.

162 Chen, Y., He, Z., Li, Y. et al. (2018). Regulation of precursor solution concentration for In-Zn oxide thin film transistors. *Current Applied Physics* 18 (11): 1300–1305.

163 He, H., Huang, H., Peng, C. et al. (2024). High-performance sol–gel processed a-IGZO TFTs with low-melting point metal electrodes. *Journal of Materials Chemistry C* 12 (2): 607–613.

164 Bae, J., Ali, A., Jang, J. et al. (2023). Spray pyrolyzed amorphous InGaZnO for high performance, self-aligned coplanar thin-film transistor backplanes. *Advanced Materials Technologies* 8 (1): 2200726.

165 Ali, A., Islam, M.M., Bae, J. et al. (2023). Low-cost and high-performance, back-channel etched InGaZnO thin film transistors by spray pyrolysis. *IEEE Electron Device Letters* 44 (10): 1632–1635.

166 Bhalerao, S.R., Lupo, D., Berger, P.R. et al. (2022). Flexible, solution-processed, indium oxide (In_2O_3) thin film transistors (TFT) and circuits for internet-of-things (IoT). *Materials Science in Semiconductor Processing* 139: 106354.

167 Isakov, I., Faber, H., Grell, M. et al. (2017). Exploring the leidenfrost effect for the deposition of high-quality In_2O_3 layers via spray pyrolysis at low temperatures and their application in high electron mobility transistors. *Advanced Functional Materials* 27 (22): 1606407.

168 Saha, J.K., Bukke, R.N., Jang, J. et al. (2020). Extremely stable, high performance Gd and Li alloyed ZnO thin film transistor by spray pyrolysis. *Advanced Electronic Materials* 6 (12): 2000594.

169 Aspoukeh, P.K., Barzinjy, A.A., Hamad, S.M. et al. (2023). A novel approach to the green synthesis of zinc oxide nanorods using *Thymus kotschyanus* plant extract: effect of ammonium hydroxide and precursor concentration. *Nano Express* 4 (4): 045001.

170 Rim, Y.S., Chen, H., Song, T.-B. et al. (2015). Hexaaqua metal complexes for low-temperature formation of fully metal oxide thin-film transistors. *Chemistry of Materials* 27 (16): 5808–5812.

171 Tsay, C.-Y., Wu, P.-H. et al. (2017). Properties of solution-processed MgInZnO semiconductor thin films and photodetectors fabricated at a low temperature using UV-assisted thermal annealing. *Ceramics International* 43 (15): 11874–11878.

172 Yu, J., Liu, G., Liu, A. et al. (2015). Solution-processed p-type copper oxide thin-film transistors fabricated by using a one-step vacuum annealing technique. *Journal of Materials Chemistry C* 3 (37): 9509–9513.

173 Su, B.-Y., Chu, S.-Y., Juang, Y.-D. et al. (2012). Improved electrical and thermal stability of solution-processed Li-doped ZnO thin-film transistors. *IEEE Transactions on Electron Devices* 59 (3): 700–704.

174 Avis, C., Jang, J. et al. (2011). High-performance solution processed oxide TFT with aluminum oxide gate dielectric fabricated by a sol–gel method. *Journal of Materials Chemistry* 21 (29): 10649–10652.

175 Avis, C., Jang, J. et al. (2022). Understanding the origin of the hysteresis of high-performance solution processed polycrystalline SnO_2 thin-film transistors and applications to circuits. *Membranes* 12 (1): 7.

176 Islam, M.M., Saha, J.K., Hasan, M.M. et al. (2021). Spray-pyrolyzed high-k zirconium-aluminum-oxide dielectric for high performance metal-oxide thin-film transistors for low power displays. *Advanced Materials Interfaces* 8 (16): 2100600.

177 Islam, M.M., Hasan, M.M., Ali, A. et al. (2022). Low cost, Al_2O_3 and $ZrAlO_x$ stack passivation by spray pyrolysis for highly stable amorphous InGaZnO thin-film transistors. *Advanced Materials Interfaces* 9 (26): 2200906.

178 Saha, J.K., Billah, M.M., Bukke, R.N. et al. (2020). Highly stable, nanocrystalline, ZnO thin-film transistor by spray pyrolysis using high-K dielectric. *IEEE Transactions on Electron Devices* 67 (3): 1021–1026.

179 Saha, J.K., Bukke, R.N., Mude, N.N. et al. (2020). Significant improvement of spray pyrolyzed ZnO thin film by precursor optimization for high mobility thin film transistors. *Scientific Reports* 10 (1): 8999.

180 Saha, J.K., Billah, M.M., Jang, J. et al. (2021). Triple-stack ZnO/AlZnO/YZnO heterojunction oxide thin-film transistors by spray pyrolysis for high mobility and excellent stability. *ACS Applied Materials & Interfaces* 13 (31): 37350–37362.

181 Ali, A., Saha, J.K., Siddik, A.B. et al. (2021). Combined effects of NH_3 and NF_3 post plasma treatment on the performance of spray coated ZnO thin film transistors. *Solid State Electronics* 186: 108178.

182 Hasan, M.M., Moyen, E., Saha, J.K. et al. (2022). Solution processed high performance perovskite quantum dots/ZnO phototransistors. *Nano Research* 15 (4): 3660–3666.

183 Saha, J.K., Bukke, R.N., Mude, N.N. et al. (2020). Remarkable stability improvement of ZnO TFT with Al_2O_3 gate insulator by yttrium passivation with spray pyrolysis. *Nanomaterials* 10 (5): 976.

184 Song, K., Yang, W., Jung, Y. et al. (2012). A solution-processed yttrium oxide gate insulator for high-performance all-solution-processed fully transparent thin film transistors. *Journal of Materials Chemistry* 22 (39): 21265–21271.

185 Avis, C., Kim, Y.G., Jang, J. et al. (2012). Solution processed hafnium oxide as a gate insulator for low-voltage oxide thin-film transistors. *Journal of Materials Chemistry* 22 (34): 17415–17420.

186 Choi, S., Park, B.-Y., Jung, H.-K. et al. (2013). Growth and characterization of sol–gel prepared Gd_2O_3 films as gate insulators for Zn–Sn–O thin film transistors. *Thin Solid Films* 534: 291–295.

187 Kim, J., Choi, S., Jo, J.-W. et al. (2018). Solution-processed lanthanum-doped Al_2O_3 gate dielectrics for high-mobility metal-oxide thin-film transistors. *Thin Solid Films* 660: 814–818.

188 Seo, J.-S., Bae, B.-S. et al. (2014). Improved electrical performance and bias stability of solution-processed active bilayer structure of indium zinc oxide based TFT. *ACS Applied Materials & Interfaces* 6 (17): 15335–15343.

189 Kim, Y.G., Bukke, R.N., Lee, J. et al. (2021). Formation of F-doped offset region for spray pyrolyzed self-aligned coplanar amorphous zinc–tin–oxide thin-film transistor by NF_3 plasma treatment. *IEEE Transactions on Electron Devices* 68 (3): 1057–1062.

190 Mude, N.N., Bukke, R.N., Saha, J.K. et al. (2019). Highly stable, solution-processed Ga-doped IZTO thin film transistor by Ar/O_2 plasma treatment. *Advanced Electronic Materials* 5 (12): 1900768.

191 Park, S.-J., Ha, T.-J. et al. (2023). Effects of Sn doping on the electrical performance and stability of sub-V operating metal-oxide thin-film transistors fabricated by oxygen annealing. *IEEE Electron Device Letters* 44 (4): 642–645.

192 Faber, H., Das, S., Lin, Y.-H. et al. (2024). Heterojunction oxide thin-film transistors with unprecedented electron mobility grown from solution. *Science Advances* 3 (3): e1602640.

193 Kim, J., Kim, T.H., Oh, S. et al. (2020). Al_2O_3-induced sub-gap doping on the IGZO channel for the detection of infrared light. *ACS Applied Electronic Materials* 2 (5): 1478–1483.

194 Kim, T.H., Park, W., Oh, S. et al. (2021). Unveiling the role of Al_2O_3 interlayer in indium–gallium–zinc–oxide transistors. *Physica Status Solidi* 218 (6): 2000621.

195 Kim, D., Yoon, S., Jeong, Y. et al. (2012). Role of adsorbed H_2O on transfer characteristics of solution-processed zinc tin oxide thin-film transistors. *Applied Physics Express* 5 (2): 021101.

196 Kim, J., Lee, J., Jang, J. et al. (2018). 80-2: AMQLED display with solution-processed oxide TFT backplane. *SID Symposium Digest of Technical Papers* 49 (1): 1080–1083.

197 An, S., Mativenga, M., Kim, Y. et al. (2014). Improvement of bias-stability in amorphous-indium-gallium-zinc-oxide thin-film transistors by using solution-processed Y_2O_3 passivation. *Applied Physics Letters* 105 (5): 053507.

198 Ji, S.-B., Seong, N., Park, J. et al. (2023). Field-effect mobility extraction of solution-processed InGaZnO thin-film transistors considering dielectric dispersion behavior of AlOx gate insulator. *ACS Applied Electronic Materials* 5 (2): 1035–1040.

199 Cho, Y.-J., Shin, J.-H., Bobade, S.M. et al. (2009). Evaluation of Y_2O_3 gate insulators for a-IGZO thin film transistors. *Thin Solid Films* 517 (14): 4115–4118.

200 Mativenga, M., Kang, D.H., Lee, U.G. et al. (2012). Study of mechanism of stress-induced threshold voltage shift and recovery in top-gate amorphous-$InGaZnO_4$ thin-film transistors with source- and drain-offsets. *Solid State Communications* 152 (18): 1739–1743.

201 Liu, H., Jing, L., He, K. et al. (2023). High performance and hysteresis-free a-IGZO thin film transistors based on spin-coated hafnium oxide gate dielectrics. *IEEE Electron Device Letters* 44 (9): 1508–1511.

202 Zeumault, A., Subramanian, V. et al. (2016). Mobility enhancement in solution-processed transparent conductive oxide TFTs due to electron donation from traps in high-k gate dielectrics. *Advanced Functional Materials* 26 (6): 955–963.

203 Kim, S., Kim, W.-Y., Im, H. et al. (2022). Effect of hydrochloric acid addition on dielectric properties of solution-processed aluminum titanium oxide thin film annealed at low temperature. *Thin Solid Films* 753: 139277.

204 Islam, M.M., Saha, J.K., Bukke, R.N. et al. (2020). Solution-processed La alloyed ZrO_x high-k dielectric for high-performance ZnO thin-film transistors. *IEEE Electron Device Letters* 41 (7): 1021–1024.

205 Yang, W., Song, K., Jung, Y. et al. (2013). Solution-deposited Zr-doped AlO_x gate dielectrics enabling high-performance flexible transparent thin film transistors. *Journal of Materials Chemistry C* 1 (27): 4275–4282.

206 Wang, B., Huang, W., Chi, L. et al. (2018). High-k gate dielectrics for emerging flexible and stretchable electronics. *Chemical Reviews* 118 (11): 5690–5754.

207 Kim, D.-G., Ryu, S.-H., Jeong, H.-J. et al. (2021). Facile and stable n+ doping process via simultaneous ultraviolet and thermal energy for coplanar ALD-IGZO thin-film transistors. *ACS Applied Electronic Materials* 3 (8): 3530–3537.

208 Chen, Q.-Z., Shi, C.-Y., Zhao, M.-J. et al. (2023). Performance of transparent indium–gallium–zinc oxide thin film transistor prepared by all plasma enhanced atomic layer deposition. *IEEE Electron Device Letters* 44 (3): 448–451.

209 Sharma, U., Kumar, G., Mishra, S. et al. (2022). Advancement of gate oxides from SiO_2 to high-k dielectrics in microprocessor and memory. *Journal of Physics Conference Series* 2267 (1): 012142.

210 Hays, D.C., Gila, B.P., Pearton, S.J. et al. (2016). Valence and conduction band offsets in sputtered $LaAlO_3/InGaZnO_4$ heterostructures. *ECS Journal of Solid State Science and Technology* 5 (12): P680.

211 Esro, M., Mazzocco, R., Vourlias, G. et al. (2015). Solution processed lanthanum aluminate gate dielectrics for use in metal oxide-based thin film transistors. *Applied Physics Letters* 106 (20): 203507.

212 Wang, W., He, G., Wang, L. et al. (2021). Solution-driven HfLaOx-based gate dielectrics for thin film transistors and unipolar inverters. *IEEE Transactions on Electron Devices* 68 (9): 4437–4443.

213 Ko, J., Kim, J., Park, S.Y. et al. (2014). Solution-processed amorphous hafnium-lanthanum oxide gate insulator for oxide thin-film transistors. *Journal of Materials Chemistry C* 2 (6): 1050–1056.

214 Chang, Y., Bukke, R.N., Kim, Y. et al. (2023). High-performance amorphous InGaSnO thin-film transistor with $ZrAlO_x$ gate insulator by spray pyrolysis. *Electronics* 12 (3): 688.

215 Hwang, S., Lee, J.H., Woo, C.H. et al. (2011). Effect of annealing temperature on the electrical performances of solution-processed InGaZnO thin film transistors. *Thin Solid Films* 519 (15): 5146–5149.

216 Jeong, S., Ha, Y.-G., Moon, J. et al. (2010). Role of gallium doping in dramatically lowering amorphous-oxide processing temperatures for solution-derived indium zinc oxide thin-film transistors. *Advanced Materials* 22 (12): 1346–1350.

217 Kim, Y.-H., Han, M.-K., Han, J.-I. et al. (2010). Effect of metallic composition on electrical properties of solution-processed indium-gallium-zinc-oxide thin-film transistors. *IEEE Transactions on Electron Devices* 57 (5): 1009–1014.

218 Nayak, P.K., Busani, T., Elamurugu, E. et al. (2010). Zinc concentration dependence study of solution processed amorphous indium gallium zinc oxide thin film transistors using high-k dielectric. *Applied Physics Letters* 97 (18): 183504.

219 Kim, G.H., Du Ahn, B., Shin, H.S. et al. (2009). Effect of indium composition ratio on solution-processed nanocrystalline InGaZnO thin film transistors. *Applied Physics Letters* 94 (23): 233501.

220 Kim, G.H., Shin, H.S., Ahn, B.D. et al. (2009). Formation mechanism of solution-processed nanocrystalline InGaZnO thin film as active channel layer in thin-film transistor. *Journal of the Electrochemical Society* 156 (1): H7.

221 Lee, W., Choi, S., Kim, J. et al. (2019). Solution-free UV-based direct surface modification of oxide films for self-patterned metal-oxide thin-film transistors. *Advanced Electronic Materials* 5 (5): 1900073.

222 Kim, G.H., Jeong, W.H., Kim, H.J. et al. (2010). Electrical characteristics of solution-processed InGaZnO thin film transistors depending on Ga concentration. *Physica Status Solidi* 207 (7): 1677–1679.

223 Hennek, J.W., Smith, J., Yan, A. et al. (2013). Oxygen "getter" effects on microstructure and carrier transport in low temperature combustion-processed a-InXZnO (X = Ga, Sc, Y, La) transistors. *Journal of the American Chemical Society* 135 (29): 10729–10741.

224 Su, B.-Y., Chu, S.-Y., Juang, Y.-D. et al. (2013). High-performance low-temperature solution-processed InGaZnO thin-film transistors via ultraviolet-ozone photo-annealing. *Applied Physics Letters* 102 (19): 192101.

225 Lim, J.H., Shim, J.H., Choi, J.H. et al. (2009). Solution-processed InGaZnO-based thin film transistors for printed electronics applications. *Applied Physics Letters* 95 (1): 012108.

226 Im, H., Ahn, J.H., Kim, Y.-S. et al. (2022). Effect of back-channel surface on reliability of solution-processed $In_{0.51}Ga_{0.15}Zn_{0.34}O$ thin-film transistors with thin active layer. *ACS Applied Materials & Interfaces* 14 (50): 55821–55827.

227 Kim, T., Kang, Y., Kim, W. et al. (2022). Optimization of dual-fuel combustion synthesis for rapid formation of solution-processed metal-oxide thin film transistors. *ACS Applied Electronic Materials* 4 (3): 1327–1334.

228 Park, S.-J., Ha, T.-J. et al. (2022). Sol-gel-based metal-oxide thin-film transistors for high-performance flexible NMOS inverters. *Journal of Alloys and Compounds* 912: 165228.

229 Zhao, X.-H., Zhuang, J., Sun, Q.-J. et al. (2023). Solution processed rare-earth doped high-k dielectrics for low-power IGZO transistors. *Journal of Alloys and Compounds* 952: 170062.

230 Park, J.W., Kang, B.H., Kim, H.J. et al. (2020). A review of low-temperature solution-processed metal oxide thin-film transistors for flexible electronics. *Advanced Functional Materials* 30 (20): 1904632.

231 Adamopoulos, G., Thomas, S., Wöbkenberg, P.H. et al. (2011). High-mobility low-voltage ZnO and Li-doped ZnO transistors based on ZrO_2 high-k dielectric grown by spray pyrolysis in ambient air. *Advanced Materials* 23 (16): 1894–1898.

232 Yu, X., Smith, J., Zhou, N. et al. (2015). Spray-combustion synthesis: efficient solution route to high-performance oxide transistors. *Proceedings of the National Academy of Sciences* 112 (11): 3217–3222.

233 Wang, B., Yu, X., Guo, P. et al. (2016). Solution-processed all-oxide transparent high-performance transistors fabricated by spray-combustion synthesis. *Advanced Electronic Materials* 2 (4): 1500427.

234 Liu, H.-Y., Hung, C.-C., Hsu, W.-C. et al. (2018). Deposition of oxide thin films by ultrasonic spray pyrolysis deposition for InGaZnO thin-film transistor applications. *IEEE Electron Device Letters* 39 (10): 1520–1523.

235 Qiu, H., Sun, H., Zheng, L. et al. (2020). Ultrasonic spray deposition of a-IGZO thin film transistor on substrate with hydrophilic patterns. *SID Symposium Digest of Technical Papers* 51 (S1): 187–189.

236 Liu, H.-Y., Chang, C.-L., Hsu, P.-H. et al. (2021). Effects of ambient/carrier gas on amorphous InGaZnO-based thin-film transistors using ultrasonic spray pyrolysis deposition. *IEEE Transactions on Electron Devices* 68 (6): 2729–2735.

237 Bae, J., Ali, A., Park, C. et al. (2023). High-performance, coplanar amorphous InGaZnO thin-film transistors by spray pyrolysis on polyimide substrate for low-cost manufacturing of foldable active-matrix organic light emitting diode display. *Journal of the Society for Information Display* 31 (5): 308–315.

238 Lim, T., Bae, J., Han, B. et al. (2023). Artificial synaptic InGaZnO thin-film transistor with long retention behavior using Al_2O_3/SiO_2 gate insulator. *IEEE Transactions on Electron Devices* 70 (1): 135–139.

239 Lee, W.H., Lee, S.J., Lim, J.A. et al. (2015). Printed In-Ga-Zn-O drop-based thin-film transistors sintered using intensely pulsed white light. *RSC Advances* 5 (96): 78655–78659.

240 Yan, X., Shi, K., Chu, X. et al. (2019). Stepped annealed inkjet-printed InGaZnO thin-film transistors. *Coatings* 9 (10): 619.

241 Secor, E.B., Smith, J., Marks, T.J. et al. (2016). High-performance inkjet-printed indium-gallium-zinc-oxide transistors enabled by embedded, chemically stable graphene electrodes. *ACS Applied Materials & Interfaces* 8 (27): 17428–17434.

242 Everaerts, K., Zeng, L., Hennek, J.W. et al. (2013). Printed indium gallium zinc oxide transistors. Self-assembled nanodielectric effects on low-temperature combustion growth and carrier mobility. *ACS Applied Materials & Interfaces* 5 (22): 11884–11893.

243 Choi, S., Kim, K.-T., Park, S.K. et al. (2019). High-mobility inkjet-printed indium-gallium-zinc-oxide thin-film transistors using Sr-doped Al_2O_3 gate dielectric. *Materials* 12 (6): 852.

244 Kim, G.H., Kim, H.S., Shin, H.S. et al. (2009). Inkjet-printed InGaZnO thin film transistor. *Thin Solid Films* 517 (14): 4007–4010.

245 Hennek, J.W., Xia, Y., Everaerts, K. et al. (2012). Reduced contact resistance in inkjet printed high-performance amorphous indium gallium zinc oxide transistors. *ACS Applied Materials & Interfaces* 4 (3): 1614–1619.

246 Wu, S., Zhang, Q., Chen, Z. et al. (2017). Inkjet printing of oxide thin film transistor arrays with small spacing with polymer-doped metal nitrate aqueous ink. *Journal of Materials Chemistry C* 5 (30): 7495–7503.

247 Zhang, Q., Shao, S., Chen, Z. et al. (2018). High-resolution inkjet-printed oxide thin-film transistors with a self-aligned fine channel bank structure. *ACS Applied Materials & Interfaces* 10 (18): 15847–15854.

248 Liang, K., Wang, Y., Shao, S. et al. (2019). High-performance metal-oxide thin-film transistors based on inkjet-printed self-confined bilayer heterojunction channels. *Journal of Materials Chemistry C* 7 (20): 6169–6177.

249 Shao, S., Liang, K., Li, X. et al. (2021). Large-area (64 × 64 array) inkjet-printed high-performance metal oxide bilayer heterojunction thin film transistors and n-metal-oxide-semiconductor (NMOS) inverters. *Journal of Materials Science and Technology* 81: 26–35.

5

Applications of Oxide Thin Film Transistors

5.1 Introduction

5.1.1 Trend in Oxide TFT Applications

In the beginning, TFTs were used as pixel-switching devices in active matrix liquid crystal displays (AMLCDs). Later, a broad range of applications in other types of displays, sensors, etc. were reported [1, 2]. Figure 5.1 shows the numbers of TFT papers applied in displays and other products in recent years summarized from Google Scholar Search. Applications beyond displays are emerging. Since TFT's electrical, optical, chemical, mechanical, magnetic, etc. properties are related to the composing material and transistor structure, all applications are restricted by these properties. For the oxide TFT, the application can take advantage of its unique device characteristics concerning the field effect mobility (μ_{eff}), threshold voltage (V_{th} or V_T or V_{TH}), on current (I_{on}), off current (I_{off}), subthreshold slope (SS), etc. Metal oxide semiconductors have high band gap energies and therefore, are transparent to the visible light.

There are three types of roles a TFT can play in a product:

(1) a control or information readout device for an attached device,
(2) a switching device in a circuit, and
(3) a sensing device for the environment.

5.1.2 Basic TFT Functions for Applications

Figure 5.2 shows the typical transfer characteristics of an a-Si:H, an a-IGZO, and a poly-Si TFT [3]. Table 5.1 lists electrical parameters extracted from these curves [3]. The electrical properties of the a-IGZO TFT are between those of the a-S:H TFT and the poly-Si TFT. Just like the a-Si:H TFT,

Oxide Thin Film Transistors, First Edition. Yue Kuo, Hideo Hosono, Michael S. Shur, and Jin Jang.

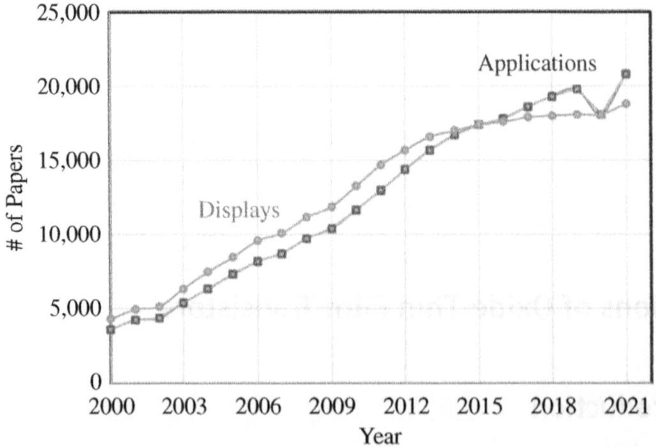

Figure 5.1 Numbers of papers on TFT displays and other applications.

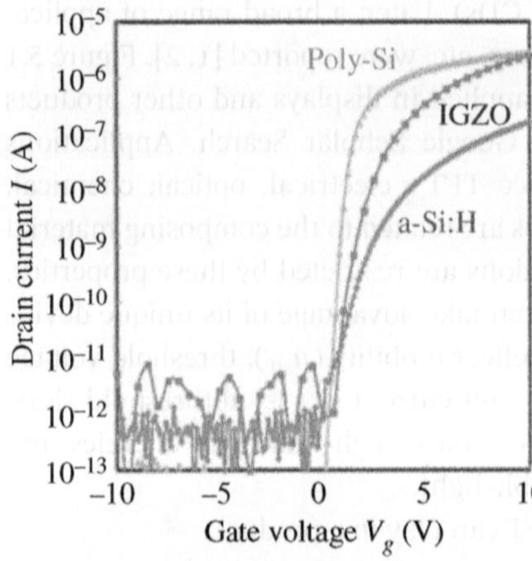

Figure 5.2 Transfer characteristics of a-Si:H, poly-Si, and a-IGZO TFTs. Source: Kuo and Chang [3]/IOP Publishing.

the a-IGZO TFT can be fabricated at a low temperature over a large-area substrate, which is difficult to achieve in the preparation of the poly-Si TFT. On the other hand, the μ_{eff} of the oxide TFT is one order of magnitude higher than that of the a-Si:H TFT. Therefore, the oxide TFT has many advantages over a-Si:H and poly-Si TFTs concerning the device performance or large-area manufacturing cost.

When a TFT is used to control or detect an attached device, there are many requirements on the electrical properties. For example, Table 5.2 lists the properties of TFTs needed to drive or monitor attached devices, such as display pixels, and optical, chemical, and gas sensors [3]. The current,

Table 5.1 Electrical parameters extracted from Figure 5.2 curves.

TFT	a-Si:H	Poly-Si	a-IGZO
W/L (μm/μm)	90/11	12/6	20/10
V_{th} (V)	1.5	0.47	1.5
μ_{eff} (cm²/V s)	0.6	235	8.9
I_{on}/I_{off}	$9.2 * 10^5$	$2.2 * 10^7$	$1.0 * 10^7$
S.S. (mV/dec)	986	224	644
References	[4–6]	[7]	[8]

Source: Kuo and Chang [3]/IOP Publishing.

Table 5.2 Electric parameters of TFTs required to drive or detect various attached devices.

Required TFT properties	Voltage	Current	Power	References
Organic light-emitting diode (OLED) pixel	15 V	10 μA	150 μW	[9]
Liquid crystal (LC) pixel	2 V	100 nA	200 nW	[10]
Photo sensor	5 V	6×10^{-10} A	3 nW	[11]
Chemical sensor (K^+)	6.1 V	1 μA	6.1 μW	[12]
Gas sensor (CO_2)	5 V	5 mA	25 mW	[13]
Gas sensor (H_2)	0.1 V	10 nA	1 nW	[14]

Source: Kuo and Chang [3]/IOP Publishing.

voltage, and power of these devices vary over a large range. In addition to the composing materials, the structure, e.g., channel width/channel length (W/L) ratio, and configuration, e.g., single- vs. dual-gate electrode or the single- vs. dual-channel layer, of the TFT affect the performance.

The TFT itself can serve as a sensor. For example, Figure 5.3 shows a diagram of electrical components in an inverted, staggered TFT. Device parameters, such as μ_{eff}, V_{th}, SS, I_{on}, and I_{off}, can change with the variation of the resistance or capacitance of a component in the structure upon exposure to light, chemicals, temperature, pressure, magnetic field, etc. The change can be quantified and correlated to the environmental parameter for sensing purposes.

In another example, the TFT can be used to control the function of an attached device. Figure 5.4a shows a diagram of a TFT connected to a resistor, which can be a sensor, an LCD, an OLED, etc. [3]. The resistance (r) of

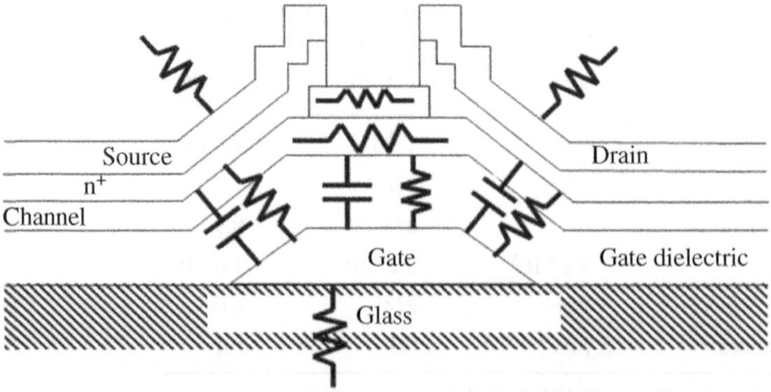

Figure 5.3 An inverted TFT with equivalent electric components of individual parts.

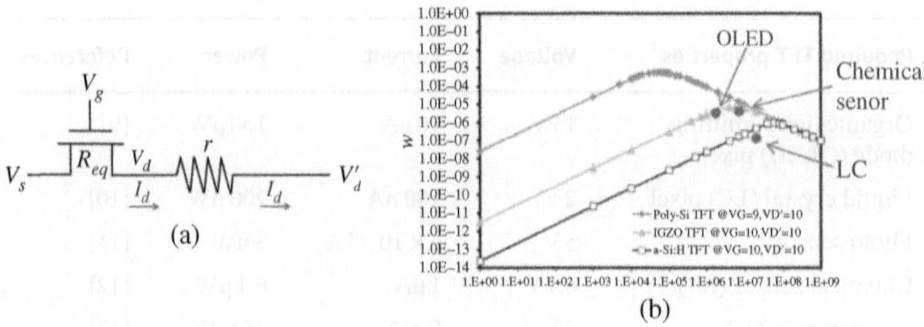

Figure 5.4 (a) A circuit diagram of a TFT with an attached resistor, (b) required TFT power vs. resistance of attached resistor. Source: Kuo and Chang [3]/IOP Publishing.

the attached device may change with the environment, operation time, and condition. Figure 5.4b shows a limit on the maximum TFT power available for the attached device. An oxide TFT is more effective than an a-Si:H TFT in driving an OLED in a pixel and certain types of sensors. For the actual product application, other factors, such as stability and reliability, need to be considered. Sometimes, a circuit including several TFTs is used in a pixel.

Figure 5.5a shows another example of a TFT attached with a resistor with resistance $r_1 = 10^7\ \Omega$ and a capacitor with capacitance $C_1 = 50\,\text{pF}$ on the gate electrode [3]. This is like the pixel structure in a touchscreen display. Figure 5.5b,c show patterns of the rise (during charge) and fall (during discharge) of the drain current (I_d, I_{ds}, or I_{DS}) with time of three types of TFTs. The response of the oxide TFT concerning time is faster than that of an a-Si:H TFT but is slower than that of the poly-Si TFT due to the mobility difference.

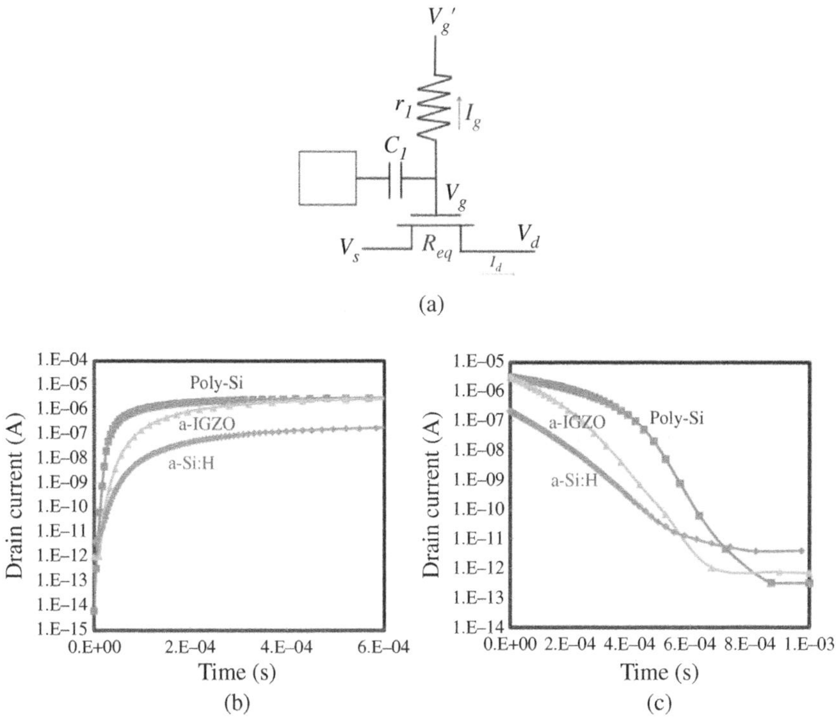

Figure 5.5 (a) A circuit diagram of a TFT attached with a resistor and a capacitor, (b) raise, (c) drop of V_d during charge and discharge, separately. Source: Kuo and Chang [3]/IOP Publishing.

Since the composing material and electrical properties of the oxide TFTs are different from those of the a-Si:H or poly-Si TFT, there are advantages and disadvantages in applications. Some of the disadvantageous properties can be improved through methods, such as the circuit or layout design and driving method, environmental control, or other means. In this chapter, examples of various oxide TFT applications are shown and discussed. Principles described in these examples are applicable to other products.

5.2 Oxide TFT Active Matrix Displays

Oxide TFTs can be used in active matrix flat panel displays (AMFPDs) to control the light transmission, emission, or reflection from an array of pixels without being interfered with by adjacent pixels.

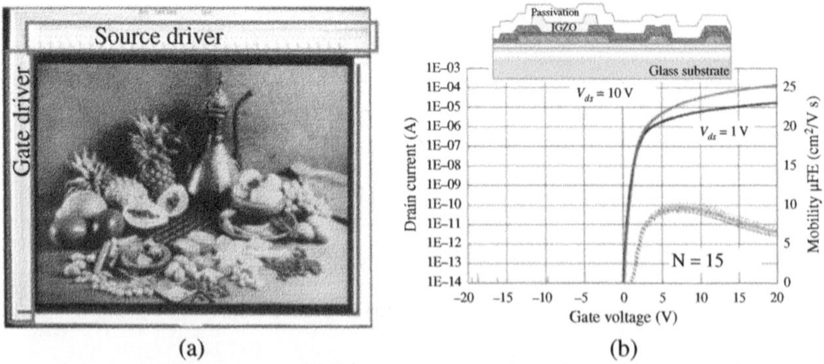

Figure 5.6 (a) A 4″ QVGA AMLCD made of a-IGZO TFTs and gate/source drivers, (b) a cross-section view and transfer characteristics of the a-IGZO TFT. Source: Osada et al. [15]/with permission of IOP Publishing.

5.2.1 Liquid Crystal Displays

Similar to the a-Si:H TFT, the oxide TFT can be fabrication at a low temperature on a large substrate. However, the high μ_{eff} of the oxide TFT enables the integration of the pixel and driver circuit process to provide high frame rate videos at a low cost.

Figure 5.6a shows a 4″ QVGA AMLCD made of a-IGZO TFTs [15]. The gate and source drivers were fabricated with the same process as that of the pixel TFT array. This display has a large aperture ratio of 41.8% and can be stably operated at a high video rate. A video signal writing period of 1.07 μs with a high source driving frequency of 234.24 kHz has been achieved. Figure 5.6b shows the coplanar TFT structure and transfer characteristics. The Ti/Al/Ti (50/100/50 nm) source/drain metal, the a-IGZO (50 nm) semiconductor, and SiO_x (300 nm) passivation layers were sputter-deposited. The SiN_x gate dielectric was deposited by plasma-enhanced chemical vapor deposition (PECVD). The complete TFT was thermally treated at 350 °C. It has a μ_{eff} of 9.65 cm²/V s and V_{th} of 1.81 V.

TFTs of top- or bottom-gate structures [16–18] and different types of oxide semiconductor materials, such as ZnO [19] or Zinc-Indium-Tin Oxide (ZITO) [16], have also been used in AMLCDs.

5.2.2 Organic Light Emitting Device Displays

Oxide TFTs can be used to control the light emission from OLEDs in the pixel array. The OLED is composed of a stacked structure of organic (hole transmission/electron-hole combination/electron transfer) thin film layers. For the rest of this chapter, OLED displays are abbreviated as OLEDs.

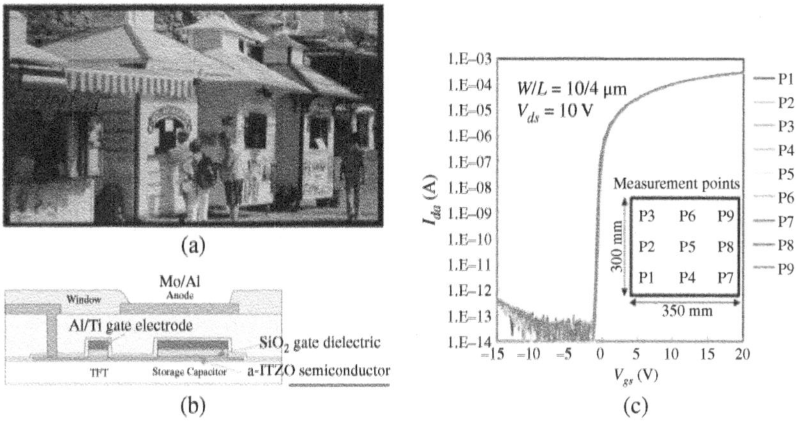

Figure 5.7 (a) A 9.9″ QHD AMOLED display panel, (b) a cross-section view of the a-IZTO TFT and storage capacitor in a pixel, (c) TFT transfer characteristics. Source: Morosawa et al. [20]/with permission of John Wiley & Sons.

Figure 5.7a shows a 9.9″ diagonal quad high resolution (QHD) AMOLED display (2,560 × 1,440 pixels) made of an Indium Zinc Tin Oxide (IZTO) TFT matrix [20]. It showed excellent bright uniformity with a peak luminescence of over 600 cd/m^2. Figure 5.7b shows the cross-sectional view of the top-gate TFT and the storage capacitor in a pixel, which were made from a 5-mask process. The IZTO film was deposited by DC sputtering, the SiO_2 gate dielectric was deposited by PECVD, the Al/Ti gate electrode stack was deposited by sputtering, and the Al_2O_3 passivation was prepared by Al sputter deposition followed by O_2 annealing, and the Mo/Al source/drain electrode stack was sputter deposited. Al was diffused into the ITZO layer in the after-deposition annealing step to form the self-aligned, low-contact resistance structure. Figure 5.7c shows uniform distribution of the transfer characteristics, with $\mu_{eff} = 32\,\text{cm}^2/\text{V s}$, $V_{th} = -0.09\,\text{V}$, and $SS = 0.\ 12\,\text{V/decade}$, across the 300 mm × 350 mm glass substrate.

Driving of an OLED pixel is more complicated than driving of an LCD pixel. Several TFTs must be included in a pixel to provide steady and uniform light emission across the whole panel. A minimum of two TFTs and one capacitor (2T1C), as shown in Figure 5.8, are required [21]. The driving TFT needs to have a large μ_{eff} and the switching TFT also must have a very low I_{off}.

Often, a relatively complicated structure is needed to provide stable and uniform light emission from an OLED pixel in a display. In-pixel and external compensation circuits have been used to provide the V_{th} stability [22, 23]. For example, Figure 5.9 shows a 3T2C layout of a pixel driving circuit in

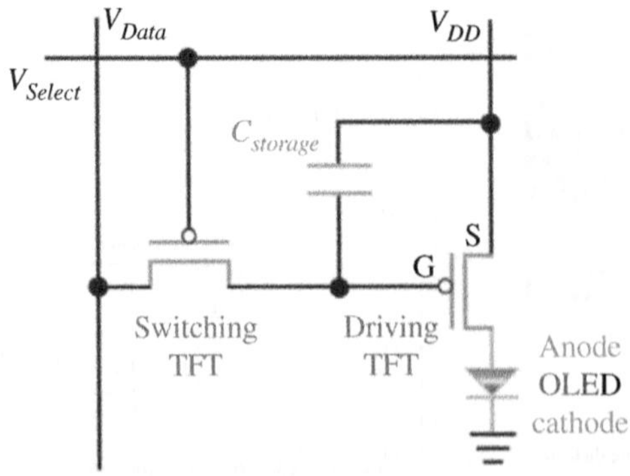

Figure 5.8 A 2T1C OLED pixel structure. Source: Zhu et al. [21]/John Wiley & Sons.

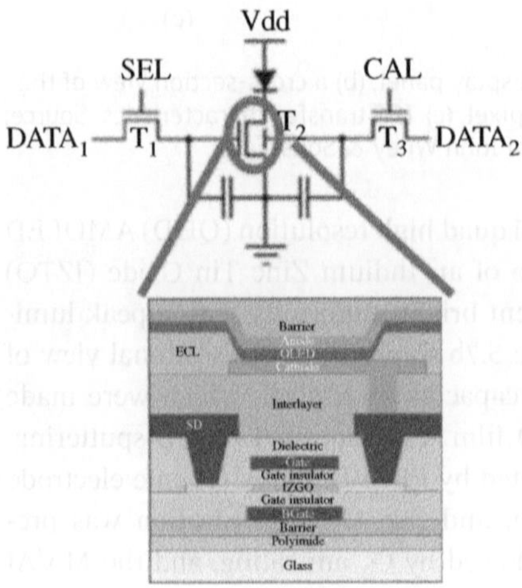

Figure 5.9 A 3T2C OLED pixel structure. Source: Verschueren et al. [22]/John Wiley & Sons.

which T2 is the drive transistor, and T1 and T3 control the data to the front and back gates, separately [22]. The T2 TFT has a dual-gate structure.

5.2.3 Micro-Light Emitting Device Displays

The OLED in the display can be replaced with inorganic GaN-based micro-sized LEDs, i.e., μ-LED displays, which will be abbreviated as μ-LEDs in the rest of this chapter. The advantages of the latter over the former are higher efficiency, lower energy consumption, higher brightness, and longer lifetime [24, 25]. The full color can be achieved by integrating blue, green, and red μ-LEDs in sub-pixels or a monochromatic blue light-emitting μ-LED with green and red color converters [26].

Figure 5.10a shows a full-color, 3.3″ a-IGZO TFT μ-LED display [27]. It is composed of an array of 232 × RGB × 116 pixels each of which is

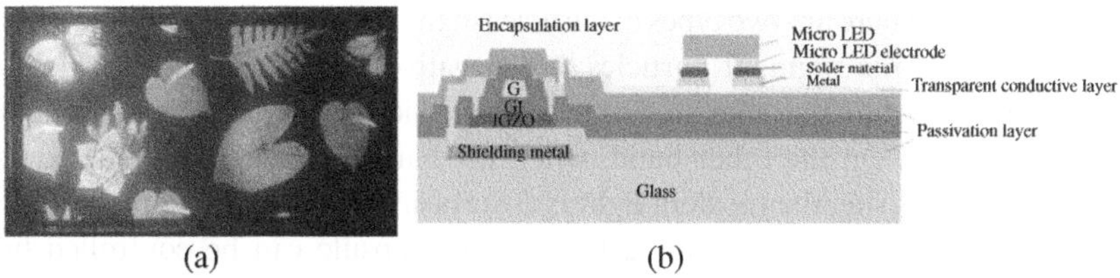

Figure 5.10 (a) A 3.3″ color a-IGZO TFT μ-LED display, (b) cross-sectional view of a pixel structure. Source: Fan et al. [27]/with permission of John Wiley & Sons.

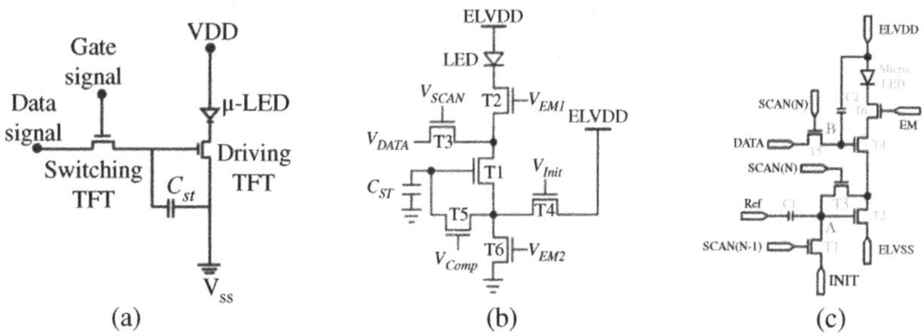

Figure 5.11 (a) 2T1C (28), (b) 6T1C [28], (c) 6T2C oxide TFT μ-LED pixel circuits. Source: Lin et al. [29]/IEEE.

324 μm × 324 μm in size. This transparent display had a high aperture of over 40%. The R, G, and B μ-LEDs with peak lights at 630, 527, and 461 nm, separately, were transferred and flip-chip bonded to individual pixels. Figure 5.10b shows the cross-sectional view of each pixel. The coplanar structured a-IGZO TFT has a bottom light shield layer connected to the data line. This display had a high 108% NTSC gamut.

Like the OLED display, the complicated multiple TFT pixel design is needed to supply the stable V_{th} shift. Figure 5.11 shows the (a) 2T1C, (b) 6T1C, and (c) 6T2C structures that have been reported for this purpose [28–30]. They can provide minimum V_{th} shift and stable current to the μ-LED, which is critical for the high-quality display. Also, the data-driving circuits can be integrated with the pixel driver fabrication process [29]. TFTs can be made of a-IGZO, a-IZO, and other types of oxide semiconductors.

5.2.4 Electrophoretic Displays

The TFT-driven electronic paper can provide repeatable, dynamic, and stable write and erasable properties with a very low power consumption. In this kind of display, microencapsulated particles are included in a thin translucent fluid layer sandwiched between a TFT plate and an ITO-coated substrate. There are two types of particles in the microcapsule, i.e., positively charged white pigment particles and negatively charged black particles. When the white particles are attracted to the viewer's site, the incident light is reflected. On the other hand, when the black particles are attracted to the viewer's site, the incident light is absorbed. The polarity and magnitude of the electric voltage applied to the microcapsule can be controlled by the TFT, which determines the intensity of the light reflected from the pixel, locally. When a color filter is added to the array, color images can be displayed. Commercial E-Ink products have been made from this type of technology containing a-Si:H TFTs [31]. When the oxide TFT replaces the a-Si:H TFT, the see-through electrophoretic display can be realized.

Figure 5.12a shows a color image of a 4″ QVGA (320 (×2) × 240 (×2)) a-IGZO TFT-driven electrophoretic display [32]. The color filter is composed of a staggered RGBW square. Figure 5.12b shows the detailed pixel

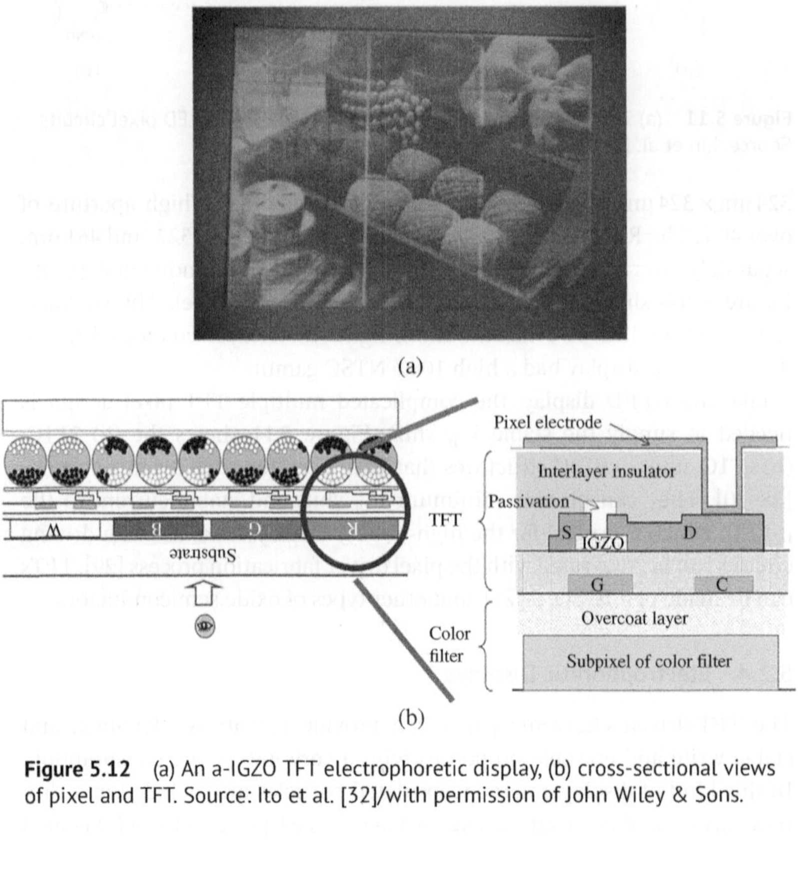

Figure 5.12 (a) An a-IGZO TFT electrophoretic display, (b) cross-sectional views of pixel and TFT. Source: Ito et al. [32]/with permission of John Wiley & Sons.

and TFT structures of the display. The color filter array was made on top of a Corning 1737 glass and subsequently coated with a transparent dielectric planarization layer. Then, TFTs were fabricated on top of the color filter substrate at a low temperature. This TFT-on-color filter structure avoids a critical alignment step, which increases the production throughput and improves reliability. The panel was finished after the TFT plate was laminated with the microencapsulate-coated imaging layer. The bottom-gate TFT ($W/L = 5/20$ μm) is composed of the sputter-deposited SiON (300 nm thick) gate dielectric, a-IGZO channel (30 nm thick), SiON (40 nm thick) passivation, gate (100 nm thick), source, and drain electrode layers. The highest temperature in the whole fabrication process was 230 °C in the polymer interlayer annealing step. The TFT had a μ_{eff} of 6.0 cm^2/V s and a V_{th} of 1.6 V, but it had a high I_{off} of $>10^{-10}$ A.

5.2.5 Other Types of Displays

Although there are few reports on other types of oxide TFT array-based displays, in principle, it is possible to use this kind of device to control an electronic or optoelectronic device in a pixel for display purposes. For example, a field emission display can be constructed by utilizing the oxide TFT to drive a micro field emission device for light emission. In this case, requirements on the TFT, such as the large driving current and high thermal budget management, must be considered.

5.3 Oxide TFT Sensors, Sensor Arrays, and Imagers

TFTs have been used as various types of sensors [33]. The basic sensing principle is the detection of the change of the device parameter from the variation of the electrical property of the TFT's bulk or interface structure. For example, a staggered a-Si:H TFT with an ion-sensitive membrane can detect the K^+ concentration in a solution sensor based on the change of the gate voltage V_g [12]. In another case, a staggered a-Si:H TFT with the Pd gate electrode can serve as the H_2 sensor in an H_2/N_2 stream based on the change of the V_{th} [34]. Yet, in another case, a poly-Si TFT with two electrodes vertically attached to the channel has been used as a magnetic sensor through detecting the Hall voltage (V_H) change [35]. The sensitivity increases with the increase of the V_d monotonically. The following are some examples of oxide TFT-based sensors.

5.3.1 pH Sensors

The solution pH sensing using Field Effect Transistor (FET) devices, such as the Ion Sensitive FET (ISFET) or Extended Gate FET (EGFET), were reported previously [36–38]. The pH sensing is based on the principle of detecting the change of transistor characteristics due to the diffusion of H^+ through a double layer formed in a solution above the gate dielectric, which can be made of SiO_2, Al_2O_3, Si_3N_4, Ta_2O_5, SnO_2, etc. Based on the same principle, oxide TFT pH sensors were demonstrated [39–41].

For example, Figure 5.13a shows the structure of an a-IGZO TFT-based pH sensor [39]. The ion-sensitive-insulator-covered bottom gate TFT is immersed in a solution. The TFT was composed of a thermal SiO_x (200 or 400 nm thickness) gate dielectric layer and an a-IGZO (50 nm) channel layer. The ion-sensitive insulator was a sputter-deposited TaO_x

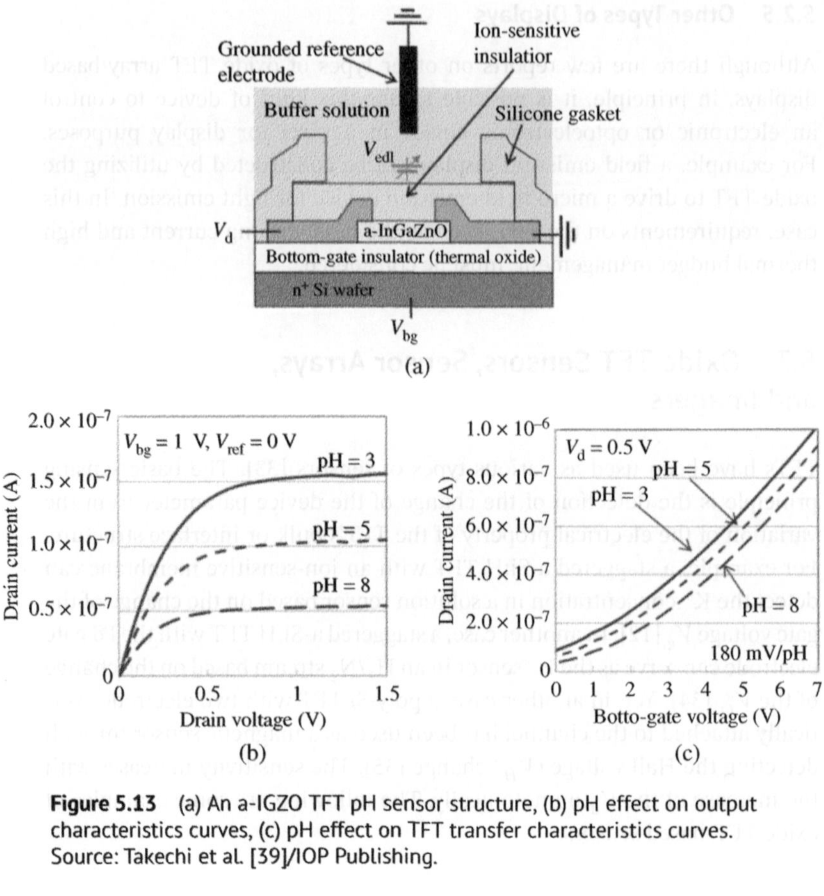

Figure 5.13 (a) An a-IGZO TFT pH sensor structure, (b) pH effect on output characteristics curves, (c) pH effect on TFT transfer characteristics curves. Source: Takechi et al. [39]/IOP Publishing.

(200 nm thick). The Ag/AgCl reference electrode (filled with KCl solution) served as a top gate electrode for the TFT. The V_{th} shift of the TFT is influenced by the ion density in the solution, e.g., the pH value, material properties of the ion-sensitive insulator, and the composition layers of the TFT, such as the dielectric constant and thickness. According to the Nernst theory, the voltage changes through the boundary layer is 59.16 mV per pH unit at 25 °C [42]. The sensitivity to pH is enhanced by the capacitive coupling of the two gate electrodes. The change of the TFT's V_{th} concerning the solution's pH value follows the equation of

$$\frac{dV_{th}}{dpH} = 59\frac{C_{IGZO}C_t}{C_b(C_{IGZO} + C_t)}(mV/pH) \tag{5.1}$$

where C_{IGZO} is the capacitance per unit of the a-IGZO layer, C_t is the capacitance per unit of the ion-sensitive insulator, and C_b is the capacitance per unit of the bottom gate dielectric layer [4].

Figure 5.13b,c are the output and transfer characteristics of the TFT, separately, in citrate/phosphate buffer solutions with the pH value varying from 3.0 to 8.0 [39]. The saturation I_d decreases but the V_{th} increases with the increase of the pH value monotonically. In a separate study using the bottom gate ZnO TFT with a Si_3N_4 passivation layer, the average current and average voltage sensitivity of 120.97 mA/pH and 97.85 mV/pH, separately, were obtained in the pH range of 3.2–11.1 [40]. A pH sensitivity as high as 359.28 mV/pH at room temperature (300 K) was achieved using a floating gate a-IGZO TFT capacitively coupled with a sensing gate (SG) and a control gate (CG) [41]. In some cases, the repeatability of the pH sensing needs improvement because the curve shifts with the operation time [39].

5.3.2 Gas Sensors

Gas sensors have been used to measure molecules of CO, CO_2, NO, NO_2, H_2S, NH_3, H_2O, etc. in the greenhouse, pollutant, biomarker, etc. gas streams. Gas sensors are also applicable to the diagnosis of food decay or diseases. The gas concentration is usually low, e.g., in the ppm range or lower. A common gas sensing method is to measure the resistance change of a resistor after the absorption of gas molecules. The reaction is dependent on the resistor's composition, porosity, morphology, temperature, etc. factors. The following are some examples of gas sensors based on oxide TFTs.

5.3.2.1 NO_2 Sensors

Figure 5.14a shows an a-IGZO TFT-based NO_2 sensor [43]. NO_2 molecules can be adsorbed onto the back-channel surface of the a-IGZO TFT. Since

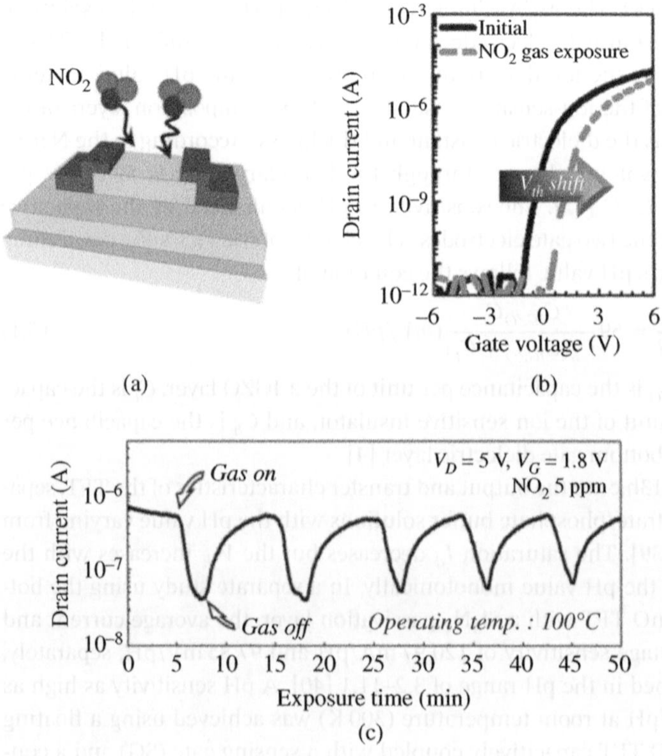

Figure 5.14 (a) an inverted, staggered a-IGZO TFT NO_2 sensor, (b) TFT transfer characteristics before and after NO_2 exposure, (c) response of I_d in response to NO_2 gas flow at 100 °C. Source: Kim et al. [43]/American Chemical Society.

NO_2 molecules are strong electron acceptors, the adsorption of them on the channel surface reduces the mobile electron density in the channel. Then, I_d is reduced and the V_{th} is shifted positively, as shown in Figure 5.14b. The reduction of I_d increases with the increase of the NO_2 concentration. The magnitude of the V_{th} shift increases with the exposure time. The same phenomena were observed in the ZnO TFT [44]. The concentration of the NO_2 gas also affects the I_d [43]. Figure 5.14c shows the response of I_d with respect to the on and off of the NO_2 (5 ppm concentration) flow at $V_d = 5$ V, $V_g = 1.8$ V, and 100 °C. The short delays in raise and fall signals, i.e., in minutes, make this TFT an effective sensor.

Temperature is important for the oxide TFT detection function because it influences the NO_2 desorption process. For example, at 293 K, the desorption of NO_2 was slow. The V_{th} remained same for a long time. However, when the temperature was increased to 470 K, the V_{th} reduced quickly upon the removal of the exposed NO_2. UV illumination is another factor affecting the detection process. For example, the illumination with a 457 nm blue light could quickly release NO_2 from the a-IGZO TFT, which resulted in the recovery of the device characteristics [45]. At the same time, the UV illumination could increase the I_{off} and SS [46]. Another method of shortening the response time of the NO_2 sensing is to pre-bias the TFT with a positive V_g. For example, when the a-IGZO TFT was pre-biased at $V_g = 5\,V$, the activation energy of the $NO_2 \rightarrow NO_2^-$ reaction on the a-IGZO surface was lowered. Then, the I_d drop rate upon the NO_2 exposure was faster than that without the pre-bias [47]. Furthermore, the NO_2 detection sensitivity of the a-IGZO TFT can be enhanced by connecting its drain electrode to the gate electrode of a poly-Si TFT [45]. The change of the I_d in the a-IGZO TFT is transformed into the change of the I_d of the poly-Si TFT. The signal could be enhanced by 3 orders of magnitude.

The same principle as that for NO_2 can be used to detect H_2S using the a-ZTO (zinc tin oxide) TFT [48]. The main difference is the direction of the V_{th} shift, i.e., toward the negative V_g direction under the H_2S exposure and the positive direction under the NO_2 exposure.

5.3.2.2 Humidity Sensors

The TFT material and structure as well as the measuring environment can be important to gas sensing [49–51]. For example, Figure 5.15a shows the transfer characteristics of an inverted, staggered a-IGZO TFT, measured in vacuum and under various humidity conditions at $V_d = 1\,V$ [49]. The original curves were measured from $V_g = -7$ to $7\,V$. The a-IGZO TFT had

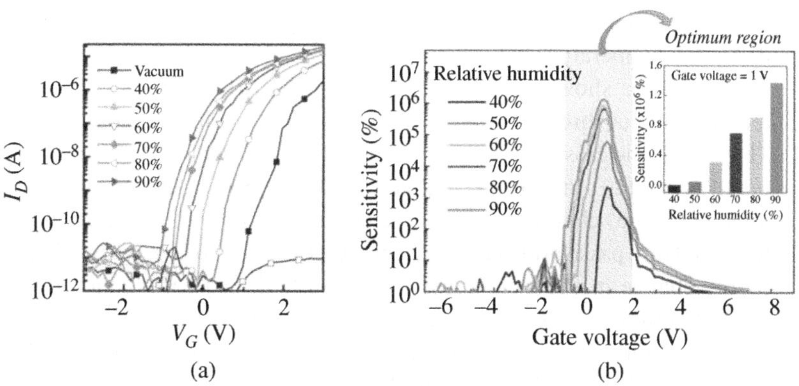

Figure 5.15 (a) Transfer characteristics of an a-IGZO TFT ($W/L = 50/500\,\mu m$) in a vacuum and various humidity conditions, (b) humidity sensitivity vs. V_g curves. Source: Kim et al. [49]/MDPI/Public domain.

an Al_2O_3 (40 nm thick) gate dielectric and ITO source, drain, and gate electrodes fabricated on top of a PVA (400 nm thick) coated glass or Si. After the PVA layer was dissolved, the TFT was transferred to a parylene (~10 μm thick) film. H_2O absorbed on the backchannel surface served as the electron donor that caused the negative V_{th} shift. H_2O absorbed on the backchannel surface served as the electron donor that caused the negative V_{th} shift. With the increase of the humidity, the V_{th} shifted from 1.55 V (in vacuum) to −0.74 V (90% relative humidity). Also, the *SS* value is little affected by the relative humidity. The low I_{off} under the humidity exposure condition indicates the lack of electrons accumulated at the backchannel surface.

Figure 5.15b shows the V_g region of the highest humidity sensitivity *S*, which is defined in the following equation. The inset shows the increase of *S* with the relative humidity. For all humidities, the highest sensitivity occurs at $V_g = -1$ to 2 V, which is near the V_{th}. The sensitivity increases with the humidity monotonically. The high sensitivity of $10^3 - 10^6$ makes this kind of TFT an effective humidity sensor.

$$S = \frac{I_{humidity} - I_{vacuum}}{W_{vacuum}} \tag{5.2}$$

5.3.3 Temperature Sensors

Temperature detection is important in many applications. It was estimated that the temperature at the center of an OLED display could be substantially raised during the operation [52]. The high temperature could generate extra oxygen vacancies in the oxide semiconductor layer of the oxide TFT to cause a change in device characteristics [53]. This phenomenon can be used to detect the temperature [53, 54].

Figure 5.16a shows that the V_{th} decreases linearly with the increase of temperature for three different a-IGZO TFTs in the range of 25–80 °C [54]. The inset is the cross-sectional of the inverted, tri-layer TFT, which is composed of Mo (20 nm)/Al (200 nm) gate electrode, SiO_2 (300 nm) dielectric, a-IGZO (50 nm) channel, and SiO_2 (200 nm) etch stop. It is covered with a SiN_x (100 nm) passivation layer. The temperature effect on V_{th} was explained by the change in the local density of states [53]. Oxygen vacancies were generated from the rise of temperature to cause the early turn on the TFT [53]. The temperature effect can be augmented by connecting several TFTs in series, as shown in Figure 5.16b. The output voltage V_{out} increased with temperature and the number of rows, but the slope did not change. Therefore, the a-IGZO TFT can be a simple temperature sensor within a defined temperature range.

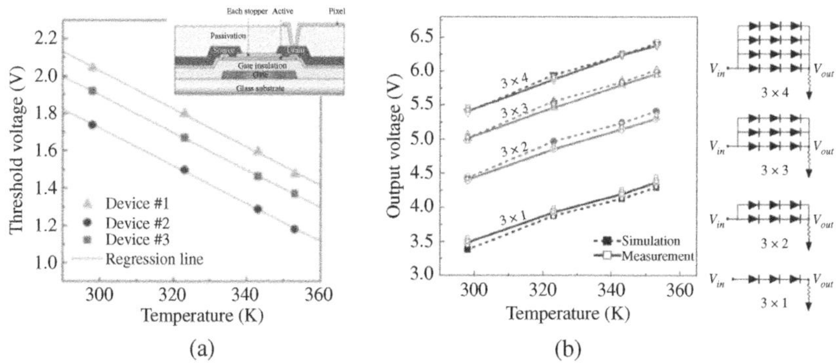

Figure 5.16 (a) V_{th} vs. temperature curves of 3 a-IGZO TFTs, inset: a cross-sectional view of the TFT, (b) V_{out} vs. temperature curves of TFTs in various arrangements. Source: Jeong et al. [54]/IEEE.

Another structure of the oxide TFT-based temperature sensor was reported [55]. The drain electrode of the a-IGZO TFT was connected to a ZnO thermistor. The output current was increased seven times when the temperature was increased from 20 to 100 °C due to the drastic decrease of the ZnO resistance in this range.

5.3.4 Pressure Sensors

Oxide TFTs can be made into pressure sensors using the piezoelectric properties of the channel material, e.g., ZnO [56, 57], or the added piezoelectric, e.g., P(VDF-TrFE)/PZT, capacitor [58]. For example, Figure 5.17a shows the cross-sectional and top views of an inverted, staggered, tri-layer ZnO TFT pressure sensor [56]. When pressure is applied to the TFT, the dipole in the ZnO channel changes, and free electrons are generated. The I_{DS} is increased accordingly. Figure 5.17b shows the quick response of the TFT characteristics with the time of the application of pressure. Also, the % increase of I_{DS} is linearly proportional to the applied pressure, as shown in Figure 5.17c. Therefore, the pressure change over a wide range can be detected from the increase of the drain current.

In another example, Figure 5.18a shows the shift of the transfer characteristics curve of a dual-gate a-IGZO TFT upon the application of various voltages to the P(VDF-TrFE)/PZT piezoelectric layer on top of the top gate electrode [58]. The inset is the cross-sectional view of the TFT and the piezoelectric layer. When a negative voltage was applied to the piezoelectric layer, the transfer characteristics shifted to the positive direction due to the establishment of the back channel. On the other hand, when a negative voltage

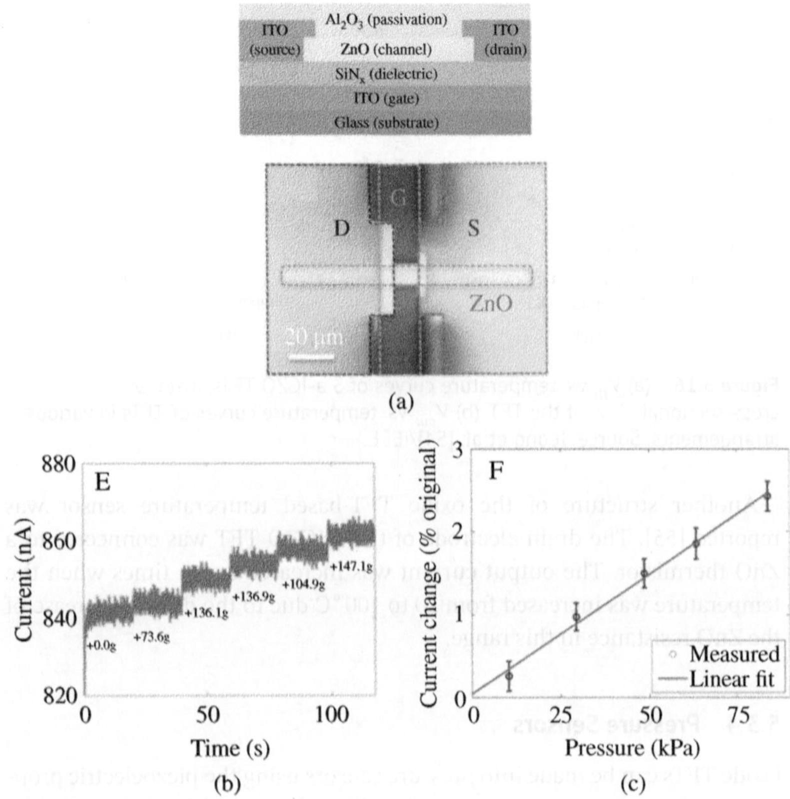

Figure 5.17 (a) Cross-sectional and top views of a ZnO TFT, (b) response of I_{DS} with pressure change, (c) I_{DS} change vs. pressure. Source: Vishniakou et al. [56]/with permission of Springer Nature.

was applied, the positive shift of V_{th} was observed due to the opposite effect. Separately, it was reported that a pressure-sensitive element, such as rubber, could be connected to the gate electrode of an a-IGZO TFT to detect the pressure change [55]. Figure 5.18b shows another design of the oxide TFT-based pressure sensor [58]. It has an inverter structure. A piezoelectric capacitor is connected to the dual gate a-IGZO TFT and a diode is connected to the single gate TFT. Upon the application of pressure to the capacitor, a voltage from the polarization of the piezoelectric material is generated. The V_{th} of the driving TFT is increased. Then, the output voltage of the sensor decreases. Once the pressure is removed, the output voltage V_{out} recovers to the original value, as shown in the inset of Figure 5.18b. The V_{out} drop is dependent on the magnitude of the applied pressure, the piezoelectric material properties, and the bias voltage applied to the dual gate TFT.

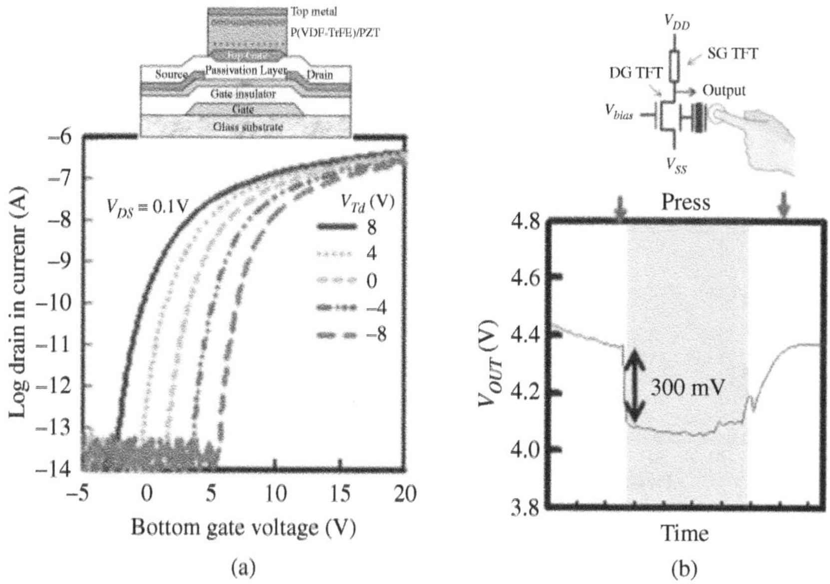

Figure 5.18 (a) Transfer characteristics vs. applied voltage, inlet: a cross-sectional view of dual-gate a-IGZO TFT/piezoelectric pressure sensor, (b) structure of an a-IGZO TFT inverter attached with a piezoelectric capacitor. Inset: V_{out} response to pressure. Source: Geng et al. [58]/IEEE.

A design of the suspended gate a-IGZO TFT pressure device has been reported [59]. Figure 5.19a shows the schematics of a dual-gate a-IGZO with a suspended top gate made of Ti (500 nm thickness) and an air gap as the gate dielectric layer. The thickness of the air gap changes with the pressure. Figure 5.19b shows the linear relationship between the displacement of the gate electrode and the pressure. Under the same electrical driving condition, i.e., fixed V_g and V_d, the I_d changes with the gate dielectric thickness. Figure 5.19c shows the relationship between the pressure and the ratio of the change of drain current (ΔI_{ds}) from the original drain voltage I_o measured at $V_g = 60$ V and $V_d = 1$ V. It shows that this kind of TFT can effectively detect the pressure above 90 kPa.

5.3.5 Bio Sensors

Conventional methods for virus detection involve complicated isolation and identification steps [60, 61]. It is time-consuming and requires special equipment. There are reports on simple, label-free biosensors that are operated based on the electrical signal change of a thin film, a transistor, a MEMS,

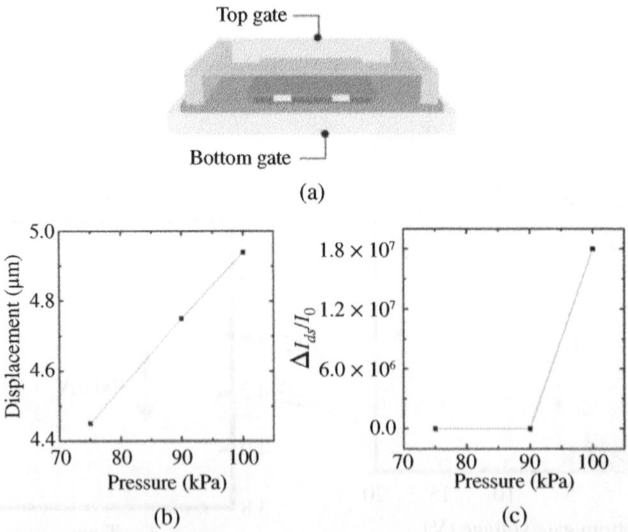

Figure 5.19 (a) Schematics of a suspended gate a-IGZO TFT, (b) displacement of gate electrode with respect to pressure, (c) $\Delta I_{ds}/I_o$ vs. pressure at $V_g = 60$ V and $V_d = 1$ V. Source: Nguyen et al. [59]/Springer Nature/CC-BY-4.0.

etc. [62–64]. Oxide TFTs can be made into biosensors because various types of reagents can be attached to them to selectively react with bio-molecules.

Figure 5.20a shows a schematic diagram of an ITO TFT virus sensor [65]. The back channel is attached with an anti-H5N1 antibody that can selectively combine with the H5N1 virus in the solution. When the antibody was not attached to the TFT, the transfer characteristics did not change upon contact with the H5N1 virus, as shown in Figure 5.20b. However, when the antibody was attached to the TFT, the transfer characteristics shifted to the positive V_g direction, i.e., increasing V_{th}, which is due to the attraction of the negatively charged virus. The magnitude of the V_{th} increase is proportional to the concentration of the virus, as shown in Figure 5.20c. The same principle can be applied to other types of viruses.

5.3.6 Photo Sensors and Imagers

5.3.6.1 UV Sensors

Like the a-Si:H TFT, the oxide TFT is sensitive to lights in the short wavelength range of visible light and UV [46, 66–69]. Figure 5.21 shows transfer characteristics of (a) an a-Si:H (30 nm thickness) TFT and (b) an a-IGZO (20 nm thickness) TFT in the dark and under the exposure of the 311 nm wavelength UV light at various intensities [68]. The maximum intensity

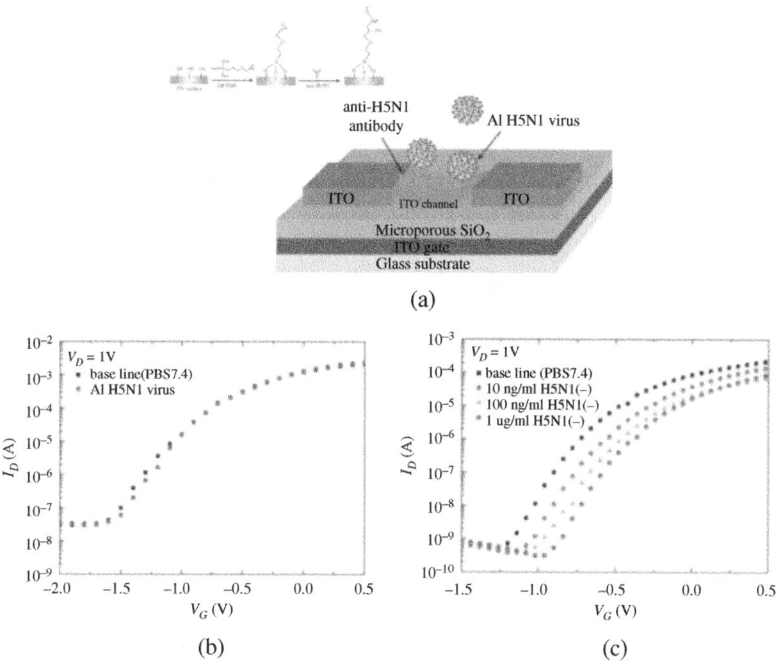

Figure 5.20 (a) Schematic of an antibody-attached ITO TFT in contact with a virus solution, (b) transfer characteristics of a TFT without antibody attachment before and after contacting virus solution, (c) transfer characteristics curves of an antibody-attached TFT before and after contacting virus solutions of various concentrations. Source: Guo et al. [65]/Elsevier.

(100%) is 1.9×10^{16} photons/cm^2 s. The increase of the leakage current and negative shift of the curve in the a-IGZO TFT are more serious than those in the a-Si:H TFT although the former has a thinner channel thickness than the latter has. The generation of sub-gap oxygen vacancy (V_0) defects and free electrons are responsible for the high light sensitivity, as shown in Figure 5.21c.

The same trend of leakage current increase and negative shift in the transfer characteristics curve under the short wavelength light exposure has been observed in other types of oxide TFTs. The magnitude of the change is dependent on the channel material and thickness, device structure, light wavelength, exposure time, gate stress voltage, temperature, etc. [46, 66–70]. For example, as shown in Figure 5.22a for the a-IGZO (50 nm thick) TFT, a large leakage current was observed when the exposure light wavelength was 400 nm or shorter [67]. Figure 5.22b shows when the channel was composed of the a-IGZO (10 nm)/amorphous indium magnesium oxide (a-IMO) (40 nm) stack, a smaller leakage current was observed at below

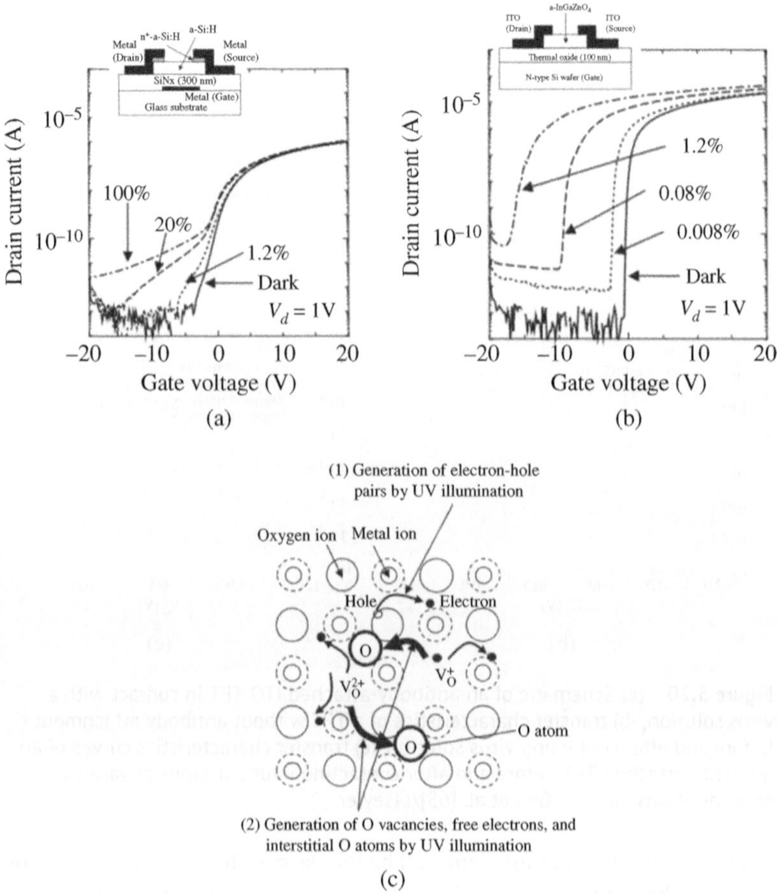

Figure 5.21 Transfer characteristics of (a) a-Si:H (30 nm thickness) TFT and (b) a-IGZO (20 nm thickness) TFT in dark and 311 nm wavelength UV lights of different intensities, (c) mechanism of UV generated defects in a-IGZO. Source: Takechi et al. [68]/IOP Publishing.

400 nm wavelength. Although the a-IMO TFT has a low mobility, the a-IGZO/a-IMO stacked TFT has a saturation mobility (μ_{sat}) close to that of the a-IGZO TFT. This can be attributed to the partial absorption of the incident light and the enhanced separation of electron-hole pairs at the interface of the two semiconductors [71].

5.3.6.2 IR and Near-IR Sensors and Imagers

Figure 5.23a shows the circuit diagram and cross-sectional view of a pixel containing an a-IGZO TFT and a perovskite diode (PD) in a flat panel

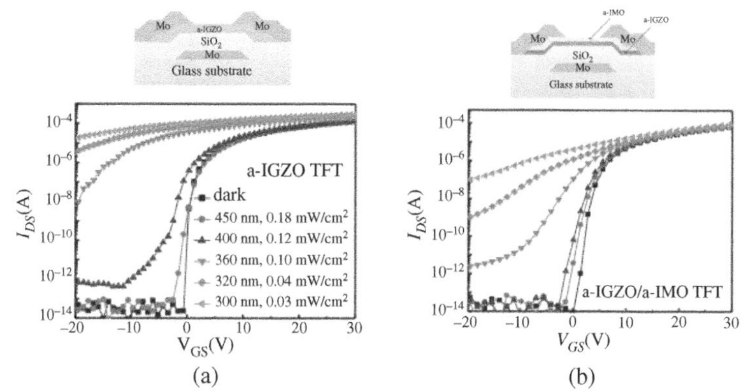

Figure 5.22 Transfer characteristics curves of (a) a-IGZO TFT, inset: cross-section view of TFT, (b) a-IGZO/a-IMO TFT exposed to lights of various lengths, inset: cross-section view of TFT. Source: Lu et al. [67]/IEEE.

near-IR sensing imager [72]. The a-IGZO TFT serves as the readout unit for the PD. A hydrophobic perfluoro(1-butenyl vinyl ether) polymer (CYTOP) layer was applied above the IGZO TFT to prevent damage from the photodiode fabrication process. The $SnO_2/(FASnI_3)_{0.6}(MAPbI_3)_{0.4}/P_3HT$ perovskite stack was spin-coated on the ITO pad and annealed. Figure 5.23b shows transfer characteristics curves of the TFT/diode stack in dark and under 850 nm light exposure at various power densities at $V_D = 0.05$ V. With the increase of the intensity of the light, the pixel current increased. Also, at the on-state of the TFT, e.g., $V_{GS} = 10$ V, the current increased linearly with the light intensity in the range of 0 – 1374 mW/cm^2 [72]. The photocurrent is limited by that of the PD not the on-current of the TFT. On the other hand, the dynamic response of the sensor is limited by the response time of the a-IGZO TFT, e.g., ~32 ms, not that of the PD, e.g., ~19 ms [72]. Figure 5.23c shows an IR (850 nm wavelength at 1374 mW/cm^2) image of "11" on a 10×10 pixel array [72].

5.3.6.3 X-ray Imagers

There are reports of the X-ray imagers based on the TFT-driven pixels [73–75]. For example, the a-Si:H pin photodiode has been used to detect the light emitted from the Gd_2O_2S:Tb or CsI:Tl scintillator under the X-ray exposure condition [74]. Although the a-Si:H TFT can be used to read out the signal from the photodiode, there are some disadvantages, such as the slow frame rate and large noises from slow signal transfer through the long data line and from the environment. When the high mobility oxide TFT is used to replace the a-Si:H TFT, these problems can be solved.

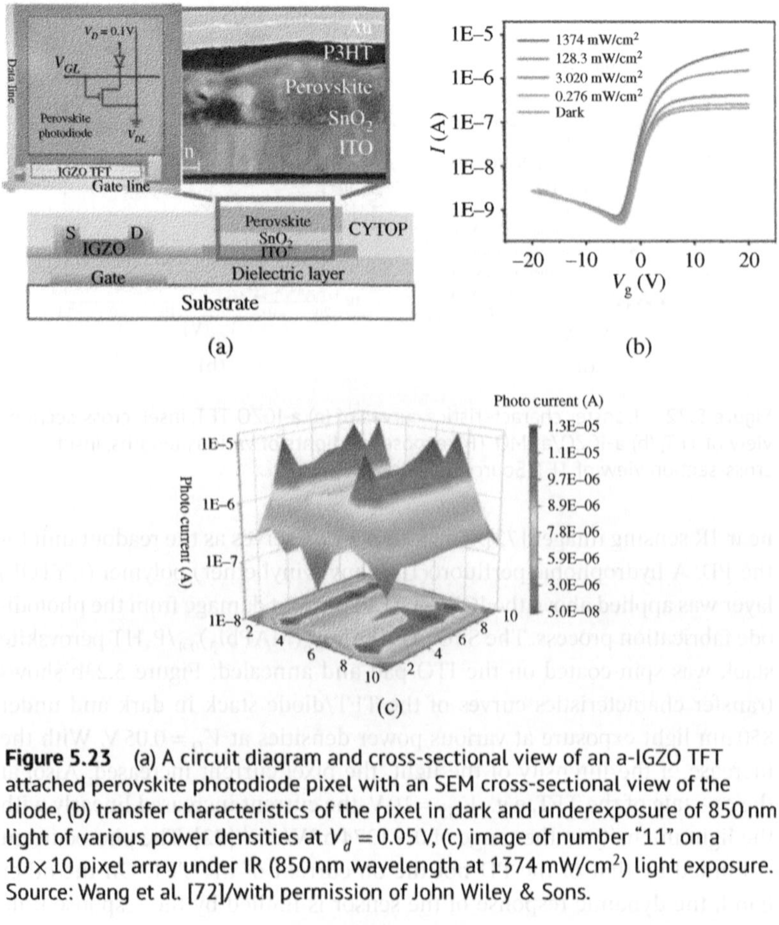

Figure 5.23 (a) A circuit diagram and cross-sectional view of an a-IGZO TFT attached perovskite photodiode pixel with an SEM cross-sectional view of the diode, (b) transfer characteristics of the pixel in dark and underexposure of 850 nm light of various power densities at $V_d = 0.05$ V, (c) image of number "11" on a 10 × 10 pixel array under IR (850 nm wavelength at 1374 mW/cm^2) light exposure. Source: Wang et al. [72]/with permission of John Wiley & Sons.

For example, Figure 5.24a shows an x-ray image of a resolution target taken from an array of a-IGZO TFT/a-Si:H pin photodiode pixels on a polyethylene naphthalate (PEN) substrate [73]. The maximum fabrication of the pixel was 170 °C. Figure 5.24b shows transfer characteristics of the a-IGZO TFT ($W/L = 100/10$ μm), which has mobility 10–15 cm^2/V s and an I_{off} of <10^{-12} A. The linear figure of the transfer characteristics is also shown. The inset is an equivalent circuit of the TFT + diode pixel. Figure 5.24c shows that the a-Si:H diode has an on (illuminated with white light)-to-off (in the dark) current ratio as high as 10^5, which is important for the high-sensitivity image.

The single TFT-driven diode design is a passive pixel sensor (PPS) that can suffer from image blurring at the high frame rate in medical

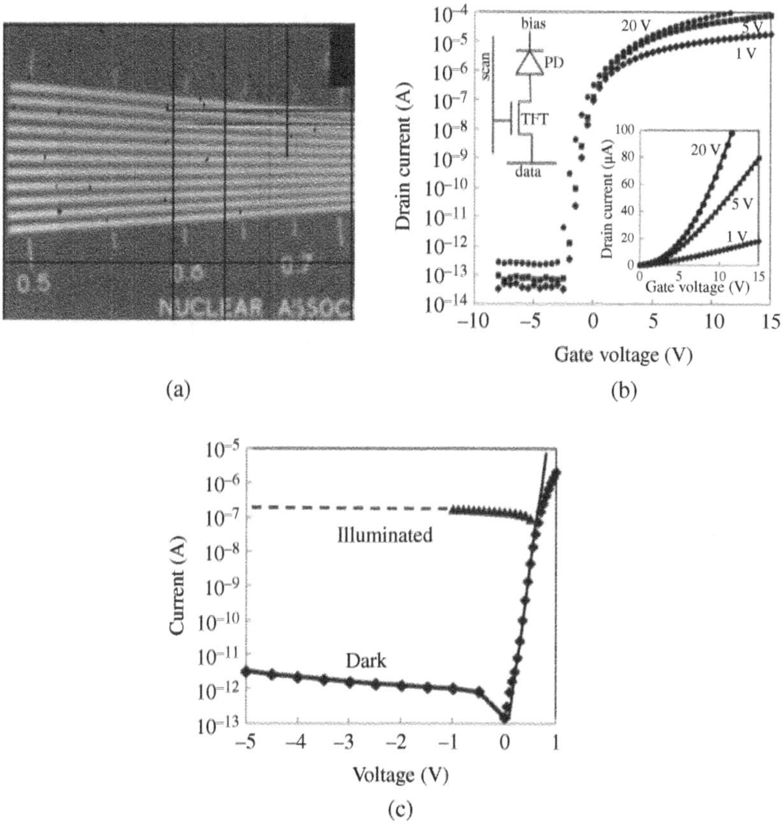

Figure 5.24 (a) X-ray image of a resolution target taken from an array of a-IGZO TFT/a-Si:H diode pixels on a PEN substrate, (b) transfer characteristics. inset: diagram of the pixel, (c) current-voltage curves in dark and under white-light illumination. Source: Lujan et al. [74]/IEEE.

diagnostic applications. The device characteristics may also change with the environment or the operation condition, such as temperature, flicker noise, external readout circuit noise, and long stress time, which causes poor image quality. The active pixel sensor (APS) architecture, i.e., including several TFTs in a pixel, can be used to stabilize or amplify the in-pixel signal quality [73, 75–79].

5.3.6.4 Oxide TFT-IC CMOS Imagers

Complementary metal-oxide-semiconductor (CMOS) image sensors (CISs) have been widely used in digital cameras for advantages, such as low power consumption. Its market is expected to grow at a compound annual growth

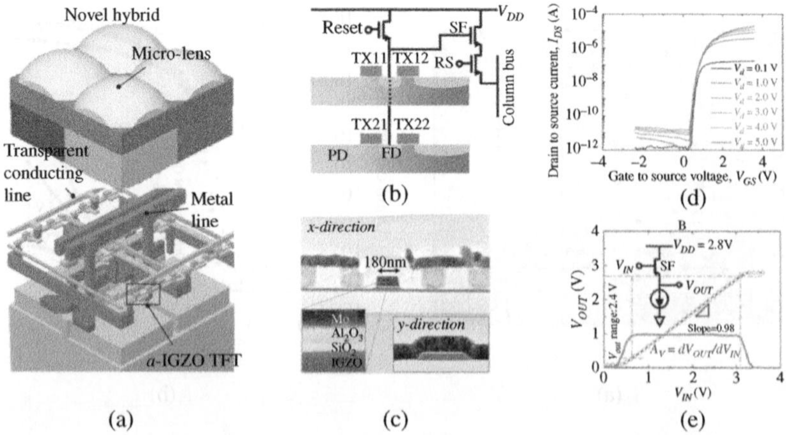

Figure 5.25 (a) A 3D view of structure and (b) equivalent circuit and cross-section drawing of a 2 × 2 pixel a-IGZO TFTs-IC hybrid CMOS image sensor unit, (c) TEM cross-sectional view and (d) transfer characteristics of the a-IGZO TFT, (e) V_{in} vs. V_{out} curve of a SF-based a-IGZO TFT/diode unit. Inset: equivalent circuit schematic. Source: Jeon et al. [77]/with permission of American Chemical Society.

rate of 10% from 2020 to 2025 [80]. The APS structure has been introduced to the CMOS imager [77]. A key issue in this application is the high resolution, which requires the minimum pixel size without losing light sensitivity.

When the oxide TFTs are fabricated on top of MOSFETs, the compact-sized CMOS imager can be realized. Figure 5.25 shows (a) a 3D view of a 2×2 pixel structured oxide TFT-IC hybrid CMOS imager with the top microlenses and (b) its equivalent circuit diagram and cross-section layout of the TFTs [76]. The RES (reset), RS (row select for readout), and SF (source follower for amplifying) are made of a-IGZO TFTs stacked on top of the interconnect lines, which provides extra area for light input to the Si photodiode. Figure 5.25c shows the TEM cross-sectional view of the Mo top gate TFT (W = 180 nm) with the SiO_2 (2.5 nm)/Al_2O_3 (5 nm) dual-layer gate dielectric. Figure 5.25d shows the transfer characteristics of the a-IGZO TFT, which has the $\mu_{eff} = 18\,cm^2/V\,s$, $I_{on}/I_{off} = 10^6$, $V_{th} = 0.6\,V$, and $SS = 70\,mV/dec$. The TFT fabrication process included Ar plasma exposure of the source and drain contact areas before metal deposition, which provided high conductivity and low contact resistance. Figure 5.25e is the V_{in} vs. V_{out} curve of the SF-based a-IGZO TFT/diode unit. The inset is the equivalent circuit. The high SF gain (A_V) of 0.98 at $V_{DD} = 2.8\,V$ is near the theoretical limit value.

To obtain the color image of a subject, 3 prime-color filters can be added on top of individual pixels. Otherwise, a three-layer-stacked color filter with organic photoconductive films (OPFs) for blue and green light detection and a Si diode for red light detection can be used [81]. For the latter case, the APS circuits can be made in the interconnect structure with the CMOS circuit made on the wafer.

An APS pixel design for blue light detection in an imager is shown in Figure 5.26 where (a) is the top view and (b) is the equivalent circuit of the pixel with an external circuit diagram [82]. The reset (M_R), amplifier (M_A), and select readout (M_S) transistors are made of indium-tin-zinc-oxide (ITZO) TFTs. The inverted, staggered structured TFT has a high field effect mobility of 24 $cm^2/V\,s$. Figure 5.26c shows the photocurrents of an OPF diode in the dark and irradiated with 465 nm blue light at a power density of 5 $\mu W/cm^2$. The diode is made of Spiro-2CBP (for electron-blocking), DPh-DNTT (for blue light sensing), and B4PYMPM (for hole-blocking) layers. The on/off ratio of $\sim 10^4$ was obtained. Figure 5.26d shows the linear output signal to the irradiation light energy density relationship. The input blue light signal can be amplified and read out easily. Based on the same principle as above, the green light signal can be read out when the blue OPF layer is replaced by a green OPF layer, such as a mixture of quinacridone (QA) and 3′,4′-dibutyl-5,5″-bis(dicyanovinyl)-2,2′:5′,2″-terthiophene (DCV3T) [81]. The highly sensitive CMOS color imager can be fabricated by stacking the blue and green APS pixels on top of the MOSFET wafer.

5.3.7 Fingerprint Sensors

Fingerprints have been used in areas requiring biometric information, such as banking, personal communications, security records, etc. Fingerprint scanners can be made of IC or TFTs, such as poly-Si TFTs [83]. When oxide TFTs are used in the latter, the fingerprint scanner can be integrated with the drivers.

One type of scanner is composed of an array of TFT-capacitor pixels. The charges stored in the capacitor change with the touch of the finger. The information is subsequently transferred to the external circuitry and the image is constructed from the distribution of charges across the whole plate [84]. Another design of the scanner is to make an array of TFT-sensor plate pixels [85]. The ridge and trough map of the fingerprint is converted to the potential of the gate that is connected in series with the perception capacitance and the parasitic capacitance of the pixel. Then, the fingerprint image is constructed by analyzing the output voltage waveform.

Fingerprint scanners can be made from oxide TFT-attached optical sensing device arrays [86] based on similar principles as those of IR, near-IR,

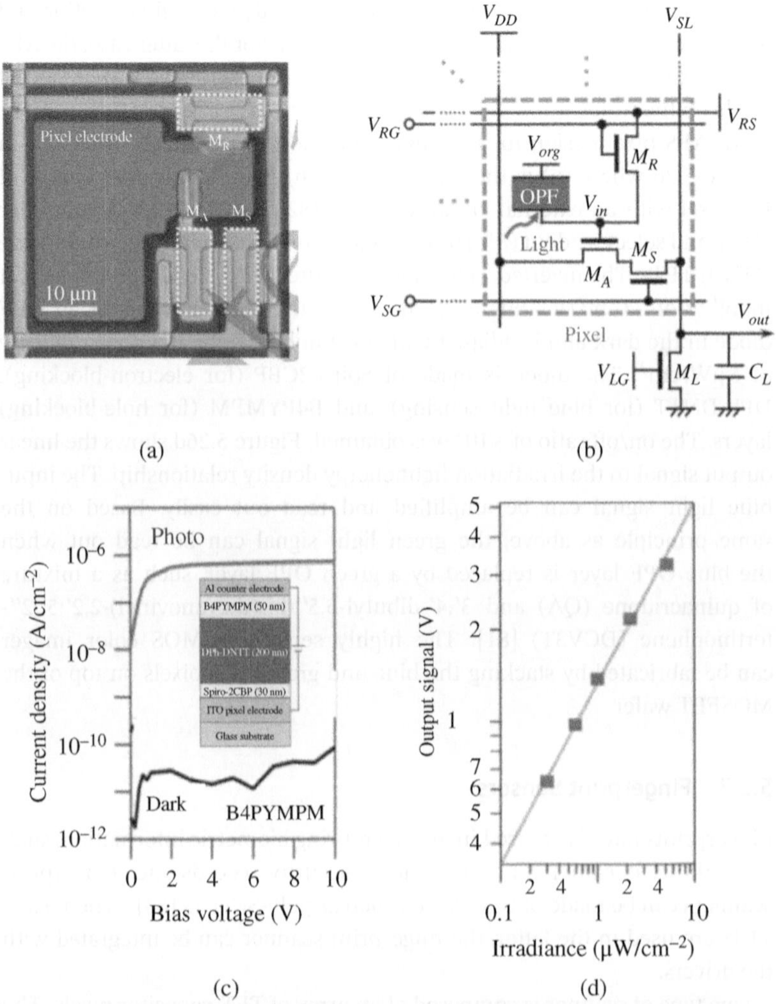

Figure 5.26 (a) Top view and (b) layout of a pixel composed of 3 a-IZTO TFTs and 1 OPF diode with an external circuit diagram, (c) current vs. bias voltage curves of a blue light photodiode measured in dark and under 465 nm blue light (at 5 μW/cm²) illumination, (d) output signal vs. irradiance power density curve. Source: Imamura et al. [82]/IOP Publishing.

or X-ray sensor arrays described in Sections 3.6.2 and 3.6.3. For example, Figure 5.27a shows the transmission and reflection of IR/near IR light from the finger to the fingerprint scanner [86]. Since the finger surface contains alleys and ridges, the light intensity received by each pixel varies locally. Figure 5.27b is the top view of a pixel composed of an a-IGZO TFT and an organic photodiode (OPD). Only a small portion of the pixel area is occupied

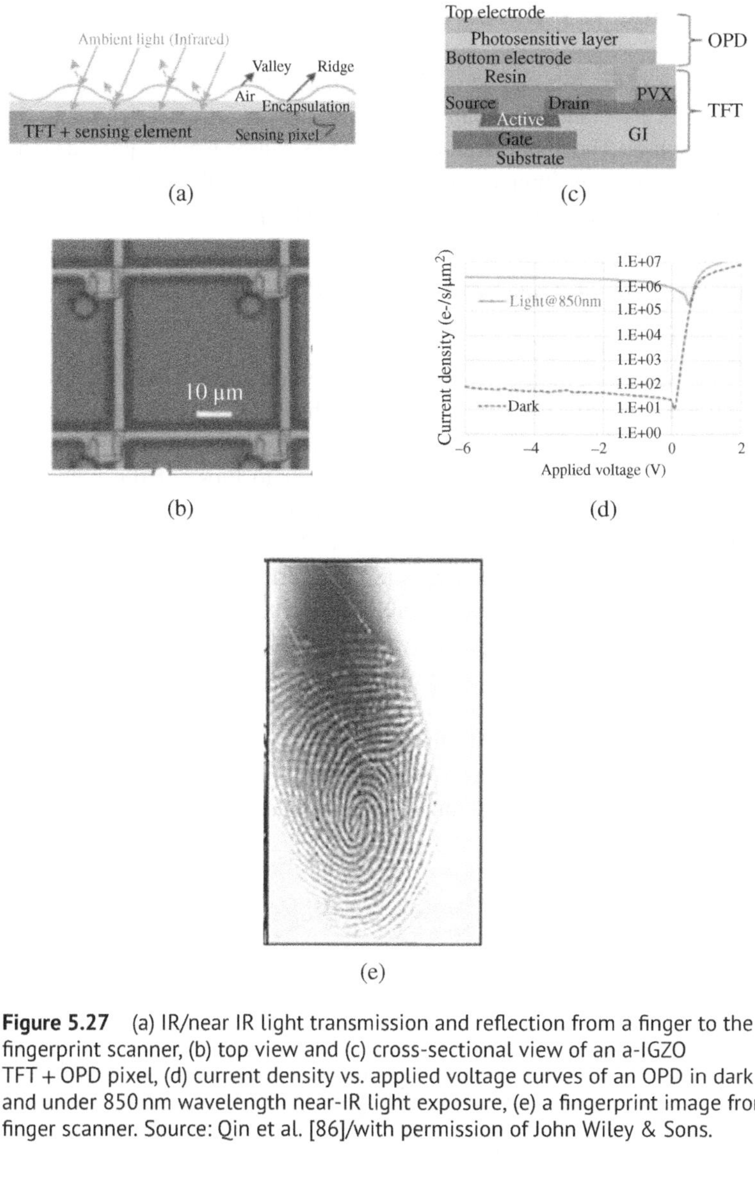

Figure 5.27 (a) IR/near IR light transmission and reflection from a finger to the fingerprint scanner, (b) top view and (c) cross-sectional view of an a-IGZO TFT + OPD pixel, (d) current density vs. applied voltage curves of an OPD in dark and under 850 nm wavelength near-IR light exposure, (e) a fingerprint image from finger scanner. Source: Qin et al. [86]/with permission of John Wiley & Sons.

by the TFT. Figure 5.27c shows the cross-sectional view of the OPD-on-TFT structure. Figure 5.27d is the current vs. voltage curves of the OPD in the dark and under the 850 nm wavelength light exposure. The current ratio of $>10^4$ between the two conditions was obtained, which is suitable for the deconvolution of the image of the finger. A high-resolution fingerprint image, as shown in Figure 5.27e, was achieved.

5.4 Oxide TFT-Based Circuits

Many types of circuits can be made using oxide TFTs. Currently, almost all oxide TFTs are made of n-channel semiconductors for pixel-driving in flat panel displays. The application of oxide TFTs can be beyond pixel driving, such as in logic and memory circuits. New types of TFTs, functions, and low-power consumption circuits are required [87].

5.4.1 Additional Oxide TFT Devices for Circuits

5.4.1.1 P-channel Oxide TFTs

For the low-power, high-performance logic circuit applications, both p- and n-channel TFTs are necessary. Although there are reports on p-channel TFTs, e.g., made of ZnO_x, CuO_x, SnO_x, etc. semiconductors [88], their performance and reliability are much inferior to those of the n-channel oxide TFTs. For example, Figure 5.28 shows the (a) output and (b) transfer characteristic curves of a bottom-gate p-channel polycrystalline SnO_x TFT [88]. The SnO_x film was deposited by thermal evaporation from a Sn target under the O_2 atmosphere followed by 100 °C, one-hour annealing. The as-deposited film was the n-type semiconductor but changed to the p-type film after the post-deposition step. Possible causes of the change of the properties are the removal of the Sn^0 state, Sn^{+2} to Sn^{+4} transformation, and inhibition of electrons from grain boundaries [87]. Like other most p-channel oxide TFTs, it has a low μ_{eff} of 0.011 cm²/V s, a small I_{on}/I_{off} of 10^3, a large SS of ~2 V/dec, and a high interface density of states (D_{it}) of

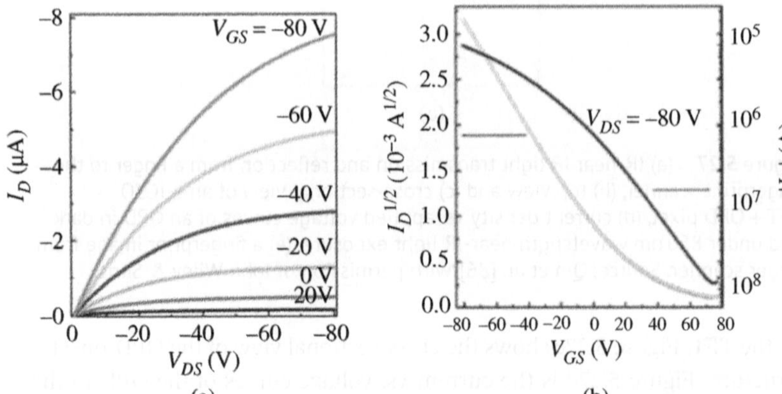

Figure 5.28 (a) Output and (b) transfer characteristics of a p-channel SnO_2 TFT, W/L = 100 μm/2000 μm. Source: Ou et al. [88]/AIP Publishing.

~8.1×10^{11} cm^{-2}. For practical applications, major improvements in device properties are required.

5.4.1.2 Floating-gate Oxide TFT Memory Devices

The floating-gate dielectric structure, i.e., a charge-trapping layer sandwiched between two dielectric layers, is commonly used in IC memories. The embedded charge-trapping layer can be a continuous film or several discrete nanodots [89]. The same structure can be used in oxide TFT memories.

Figure 5.29a shows the band gap diagram of the gate dielectric/channel region in a floating-gate a-IGZO TFT [90]. The gate dielectric was composed of an Al_2O_3 tunneling layer (TL) /Si_3N_4 charge trapping layer (CTL)/SiO_2 blocking layer (BL). Figure 5.29b shows the positive shift of the transfer characteristic curve upon the application of a positive V_g pulse due to the

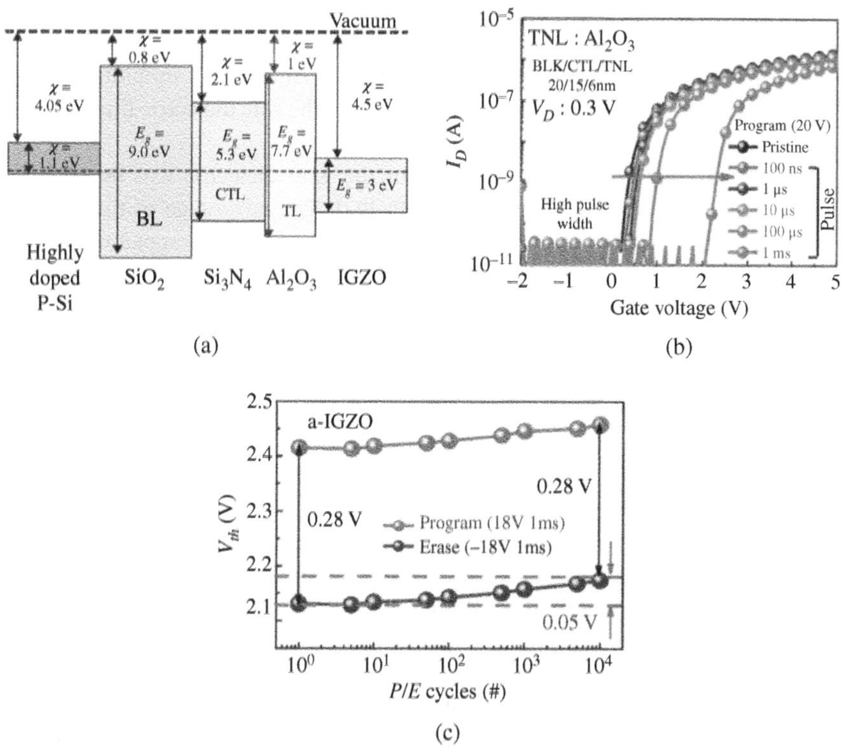

Figure 5.29 (a) A band diagram profile of the (SiO_2/Si_3N_4/Al_2O_3) floating-gate/a-IGZO channel region, (b) transfer characteristics under different V_g pulse time, (c) memory window vs. *P/E* cycles. Source: Han et al. [90]/MDPI/Public domain.

injection and trapping of electrons in the gate dielectric layer. The increase of the V_{th} is dependent on the number of electrons trapped, which is proportional to the magnitude and V_g stress time. The trapped electrons can be erased upon the application of a large negative V_g. The memory window is determined by the V_{th} difference between the program state, e.g., after V_g stress, and the erase state, e.g., after $-V_g$ stress. Figure 5.29c shows that the V_{th} in the program or erase states vs. the program/erase (P/E) cycle number. An increase of 0.05 V in V_{th} in both program and erase curves after 10^4 cycles has been observed but a gap of 0.28 V between these two curves, i.e., the memory window, was maintained.

The memory window is dependent on factors, such as the channel material, gate dielectric materials and thicknesses, magnitude of stress V_g, and environment, such as temperature and light exposure. For example, Al_2O_3 is a better TL material than SiO_2 because it has a smaller energy band gap, i.e., a smaller barrier height for electron transfer, and better electron affinity [90]. On the other hand, with the same floating-gate dielectric, the c-axis aligned crystalline IGZO (CAAC-IGZO) TFT had a larger memory window than that of the a-IGZO TFT because the former's high mobility enabled better injection and retention of electrons in the gate dielectric film than the latter [91].

Separately, the embedded layer in the gate dielectric can be replaced by discrete nanoparticles, such as Pt [92]. The advantage of this kind of gate dielectric structure is the prevention of losing all trapped charges in the embedded layer due to one leaking path in the thin tunnel dielectric layer [89].

5.4.1.3 Ferroelectric-gate Dielectric Oxide TFT Memory Devices

When an electric field is applied across a ferroelectric or piezoelectric dielectric thin film, the polarity of functional groups changes. For example, when a MIM capacitor with a $Bi_{4-x}La_xTi_3O_{12}$ (BLT) ferroelectric film sandwiched between two Pt electrodes is applied with cycles of V_g's of oppositive polarities, the polarization-electric field (P-E) hysteresis is observed due to the storage and release of charges [93]. The number of stored charges is determined by the magnitude and polarity of the electrical field. On the other hand, when the non-ferroelectric SiO_2 or HfO_2 was used in the capacitor, no hysteresis was observed.

When an oxide TFT contains a ferroelectric gate dielectric, the channel conductance changes upon the switching of the polarity of the V_g. For example, Figure 5.30 shows the hysteresis transfer characteristics of a Sb-doped SnO_2 TFT with a $PbZr_{0.2}Ti_{0.8}O_3$ ferroelectric gate dielectric when the V_g was swept from −2.5 to +2.5 V (forward) and back to −2.5 V

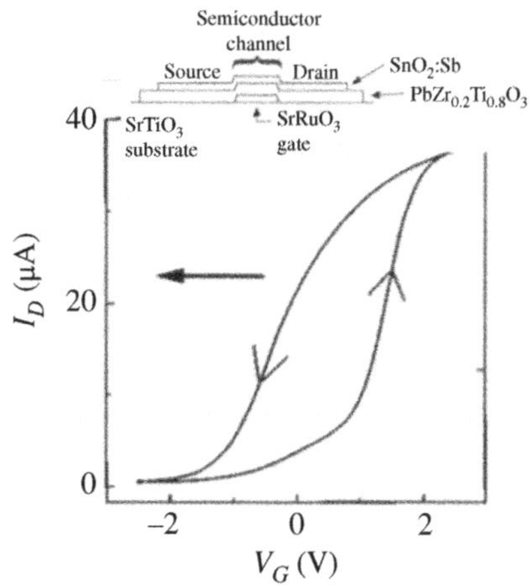

Figure 5.30 Structure and transfer characteristics of an Sb-doped SnO_2 TFT with $PbZr_{0.2}Ti_{0.8}O_3$ gate dielectric $W/L = 300/5\ \mu m$. Source: Prins et al. [94]/AIP Publishing.

(backward) [94]. The counterclockwise hysteresis was caused by the charge and discharged states of the gate dielectric film. Since the polarity change can last for a long period, this is a nonvolatile memory device. A similar hysteresis phenomenon can be observed in the floating-gate TFT, which is caused by a different mechanism, i.e., charge injection into and de-trapping in the gate dielectric film.

Both inorganic ferroelectrics, such as BLT, PZT ($PbZr_{0.2}Ti_{0.8}O_3$ or $PbZr_{0.4}Ti_{0.6}O_3$), Si-doped HfO_2 [93–97], and organic ferroelectrics, such as P(VDF-TrFE), have been used in TFT memories [98]. Figure 5.31a shows the cross-sectional view of an a-IGZO TFT with the Al_2O_3/P(VDF-TrFE) ferroelectric gate dielectric [98]. Figure 5.31b shows the transfer characteristic and gate leakage current I_g curves of the TFT with V_g swept from −6 to +6 V and then back to −6V. The low I_g is attributed to the existence of

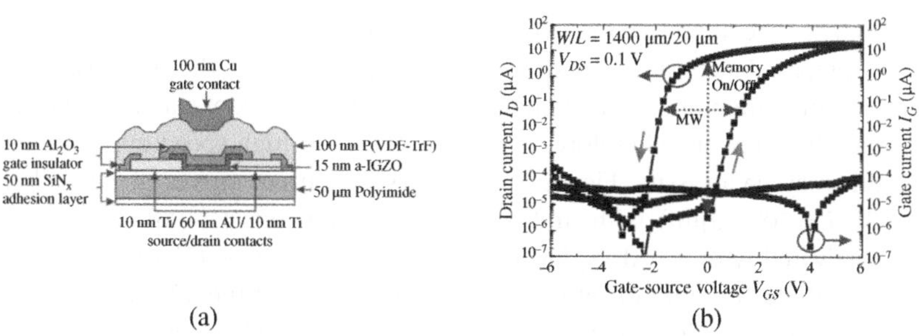

Figure 5.31 (a) Cross-section view of an a-IGZO TFT with Al_2O_3/p(VDF-TrFE) gate dielectric and (b) transfer characteristics and I_g hysteresis curves with V_g swept from −6 to +6 V then back to −6 V. Source: Petti et al. [98]/IEEE.

the Al_2O_3 layer. The large hysteresis of the transfer characteristic curve, i.e., V_{th} shift (ΔV_{th}) of 3.2 V, comes from the polarization of the p(VDF-TrFE) layer. The memory function can also be shown from the large I_{on} (I_D of the backward curve)/I_{off} (I_D of the forward curve) ratio of 1.5×10^6 at $V_g = 0$ V.

TFT ferroelectric memories can be made into an array form on a glass or plastic substrate. For example, Figure 5.32a shows a memory array circuit [99]. Each pixel contains one datum-storage memory a-IGZO TFT (FT) with an Al_2O_3/P(VDF-TrFE) ferroelectric gate dielectric and two regular a-IGZO TFTs (TA and TB) to control the write, hold, and readout functions of the FT. Transfer characteristics of these two kinds of TFTs are shown in the inset. The regular TFT did not show hysteresis while the ferroelectric Gare dielectric TFT showed obvious hysteresis. Figure 5.32b shows the readout currents for the basic write/readout operation in a pixel [99]. With this design, a write/read current ratio higher than 10^4 was achieved. Separately, it was proved that the array could be operated disturb-free, i.e., accessing a pixel without affecting data stored in other pixels. Also, the data were not damaged during the readout process.

5.4.1.4 High Voltage Oxide TFTs

Conventional TFTs are operated at low voltage, e.g., for displays. For applications in X-ray imaginers, MEMS, field emitter arrays, actuators, etc., high operation voltages, e.g., several hundreds or thousands of volts between source and drain, are necessary [100–104]. New composing materials or structures are needed.

There are many reports on high-voltage a-Si:H and poly-Si TFTs [104–108]. The common approach is to offset the source and/or drain from the gate electrode. During operation, the gated and offset regions are depleted, which reduces the lateral electric field. A large source-to-drain voltage is required for the current to flow. The high-voltage oxide TFT can be prepared using a similar structure. Figure 5.33a shows the cross-sectional view of a drain offset high voltage TFT. Due to the lack of gate influence in the offset region. Therefore, the V_{th} is high. For high-speed applications, this may be an issue. Figure 5.33b shows the output curves of ZnO TFTs ($W/L = 200/20\,\mu m$) with and without the drain offset [106]. Both were passivated with an Al_2O_3 layer. The breakdown voltage (V_{BD}) of the TFT with the 2 μm drain offset is larger than that without the offset, e.g., 82 vs. 34 V. The breakdown happened at the drain region where the electric field change was the largest. In addition, for the drain-offset TFT, the I_{ds} in the saturation region increased with the increase of the V_{ds}. The V_{th} shifted due to the self-heating effect, which was difficult to dissipate through the

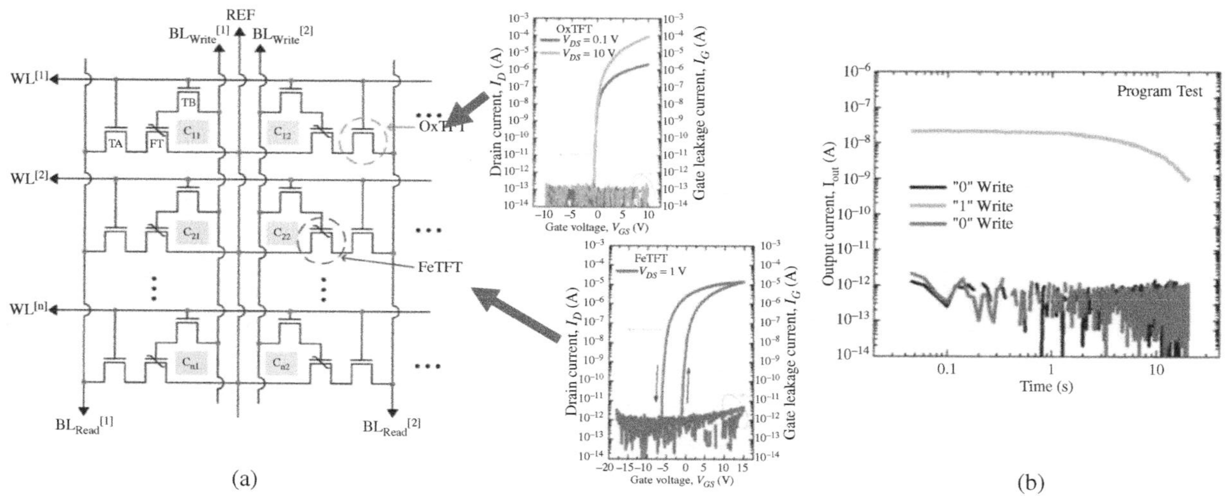

Figure 5.32 (a) A memory array circuit diagram, insets: transfer characteristics of an a-IGZO TFT and a ferroelectric memory TFT, (b) readout currents for basic write/readout operations. Source: Kim et al. [99]/IEEE.

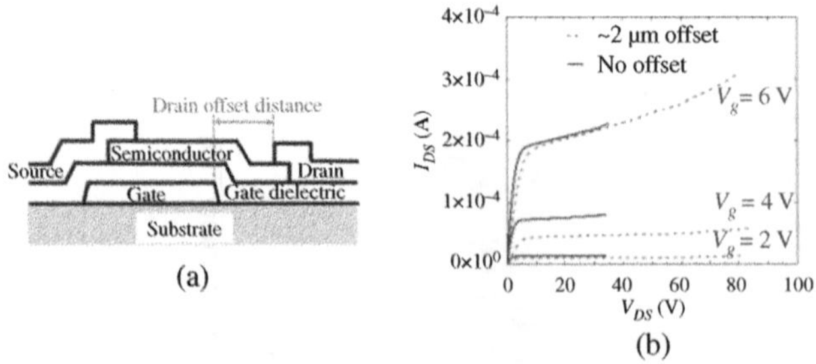

Figure 5.33 (a) A Cross-sectional view of a drain-offset TFT, (b) output curves of ZnO TFTs ($W/L = 200/20$ μm) without and with a 2 μm drain-to-gate offset. Source: Gong and Jackson [106]/IEEE.

low thermal conductivity glass or plastic substrate [109]. The self-heating can be reduced using the pulse mode gate driving method [106].

It was reported that a staggered YZTO (Yttrium Zin Tin Oxide) TFT ($W/L = 5/500$ μm) with the drain-to-gate offset of 50 μm could be operated at $V_{ds} = 500$ V [110]. In another case, an inverted, staggered a-IGZO TFT ($W/L = 60/200$ μm) with the drain-to-gate offset of 100 μm had a large V_{BD} of 1,100 V [111].

Another TFT structure has been used for the high-voltage operation. A field plate (FP) was added to the back channel of the drain offset TFT to further increase the operation voltage, i.e., enlarging the V_{BD} [102, 104]. The field plate can be located on the left (left-FP), the edge (middle-FP), or right (right-FP) the gate electrode edge. Figure 5.34a shows a drain offset IGZO TFT ($W/L = 100/100$ μm, offset length L_{offset} 20 μm) with a FP (length L_{FP} 5 μm) located at a distance X_{FP} from the edge of the source electrode.

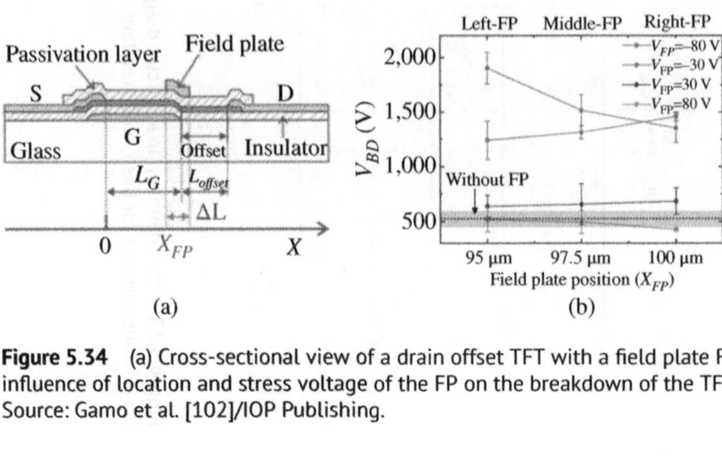

Figure 5.34 (a) Cross-sectional view of a drain offset TFT with a field plate FP, (b) influence of location and stress voltage of the FP on the breakdown of the TFT. Source: Gamo et al. [102]/IOP Publishing.

Figure 5.34b shows that the V_{BD} of the TFT was increased with the addition of the FP. The magnitude is dependent on its location, magnitude, and polarity [102]. A very large V_{BD} of 1,400 V was obtained when the FP was located at the top of the gate electrode and stressed at −80 V. In the drain offset TFT with the FP negatively biased, the peak electric field shifted away from the edge of the gate.

5.4.2 Inverters and Ring Oscillators

For circuit applications, both the DC and dynamic characteristics must be measured [112, 113]. The former is determined from the inverter performance; the latter could be judged using the ring oscillator. Figure 5.35 shows 2 layouts of inverter structures: (a) TFTs made of the same n-channel material and (b) TFTs made of a n-type and a p-type channel material. Inverters can also be made of TFTs of different layouts, such as connecting the gate of the load TFT to V_{DD} [115]. Figure 5.35c is the voltage transfer characteristics (VTC) of the inverter [114]. The magnitude of the output voltage (V_{OUT}) changed from a steady high value to a steady low value in a small input voltage (V_{IN}). The high Gain, which is defined as (dV_{OUT}/dV_{IN}), is necessary for the sustained signal propagation in an integrated circuit comprising many stages.

The inverter can be made of 2 n-channel TFTs. For example, Figure 5.36a shows the output characteristics of an inverter composed of 2 amorphous indium gallium oxide (a-IGO) TFTs [112]. The inset is the gain vs. V_{DD} curve. Figure 5.36b shows the linear relationship between the operation frequency and the V_{DD} of a 5-stage ring oscillator made of a-IGO TFTs [112]. The inset shows the oscillation of the V_{OUT} with the time. The maximum

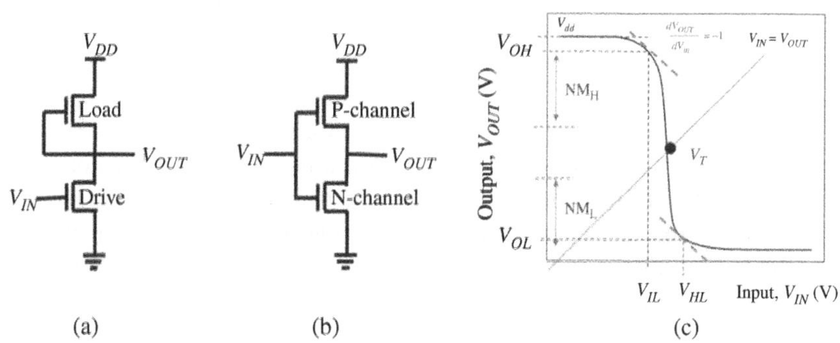

Figure 5.35 Layouts of inverters made of 2 TFTs (a) of the same type of channel material, (b) of different types of channel materials, and (c) output characteristics of an inverter. Source: Nomura [114]/Taylor & Francis Group/CC BY 4.0.

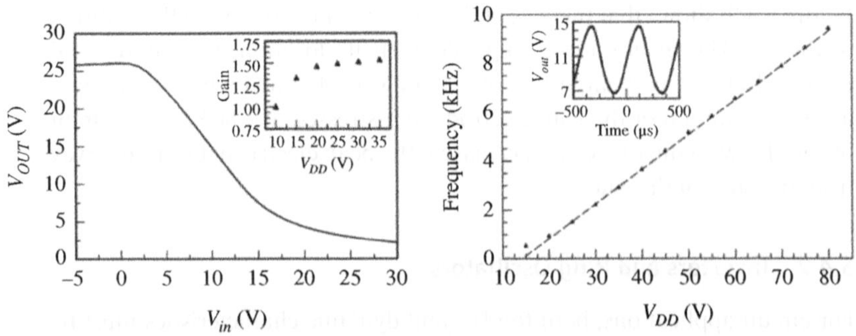

Figure 5.36 (a) Transfer curve of an inverter made of 2 a-IGO TFTs, inset: inverter gain vs. V_{DD}, (b) oscillation frequency of a 5-ring oscillator vs. V_{DD}, inset: output characteristics at $V_{DD} = V_{load} = 30$ V. Source: Presley et al. [112]/Elsevier.

oscillation frequency of 9.5 kHz at $V_{DD} = 80$ V was obtained. The maximum oscillation frequency can be increased by improving the TFT characteristics. For example, a high oscillation frequency of 410 kHz was obtained at a small voltage of +18 V when the mobility of the TFT was improved, and the gate-to-drain and gate-to-source parasitic capacitances were reduced. A similar VTC has been obtained when SnO_2 p-channel TFTs are used in the inverter [88].

There are many efforts to improve the mobility of the oxide TFT. For example, the polycrystalline In_2O_3 doped GaO_2 (poly-IGO) TFT with a high μ_{eff} of 51.4 cm^2/V s was prepared. Figure 5.37a shows the transfer characteristics of the poly-IGO TFT [115]. The poly-IGO film, as shown in the top inset XRD pattern, was formed by 400 °C annealing of the sputter-deposited film. The low I_{off} of 10^{-13} A was attributed to the source- and drain-to-gate offset structure, which is also shown in the inset. Figure 5.37b shows the output characteristics of a 23-stage ring oscillator made of poly-IGO TFTs. The top inset shows the structures of the single-stage inverter and the 23-stage ring oscillator. The high oscillating frequency of 2.36 MHz with a propagation delay time of 9.2 nm/stage at a supply voltage of 20 V was achieved. It is much higher than that of the a-IGO TFT ring oscillator.

It is difficult to prepare the CMOS-type oxide TFT circuit because the p-channel oxide TFT generally suffers from low μ_{eff}, large I_{off}, and poor reliability. It can be solved using the p-channel low-temperature poly-Si TFT (LTPS) or organic TFT with the n-channel oxide TFT [116–120]. The LTPS and oxide TFTs combination is called LTPO (low-temperature poly-Si and oxide TFTs).

Figure 5.38 shows (a) the cross-sectional and (b) the top views of a LTPO inverter on a glass substrate [119]. The poly-TFT was made of the

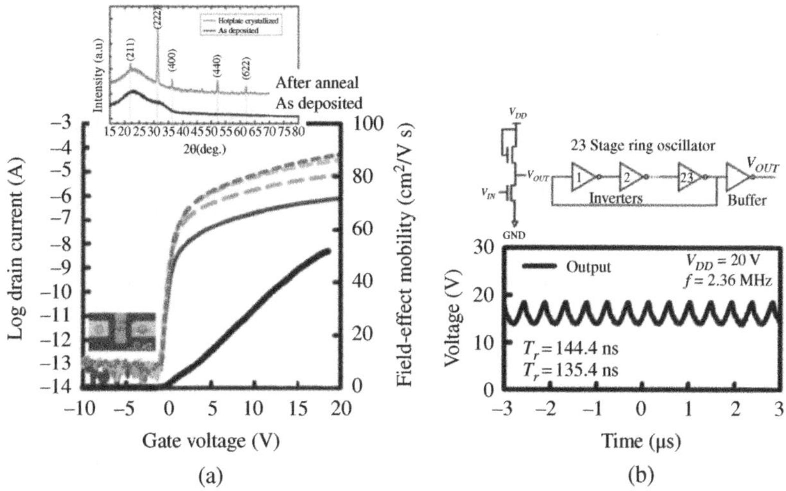

Figure 5.37 (a) Transfer characteristics of a poly-IGO TFT, top inset: XRD of poly-IGO, bottom left inset: source- and drain-offset structure, (b) output characteristics of a 23-stage ring oscillator, top left inset: inverter structure, top right inset: ring oscillator layout. Source: Rabbi et al. [115]/John Wiley & Sons.

excimer laser annealing (ELA) crystallized a-Si. The 2 TFTs are fabricated side-by-side sharing the same gate driving voltage. Figure 5.38c shows the VTC and the crossing current-voltage curves of the inverter. The inset is the circuit diagram of the inverter. A large inverter gain of 68.3 and a threshold voltage V_{IN} close to half of the V_{DD}, e.g., 4.98 vs. 5 V, were achieved. It was also stable against the noise signal due to the large symmetry margin ($NM_H = 4.12$ V and $NM_L = 4.53$ V). It had a very low power dissipation at the steady state because of the low crossing current. With a similar design, an even higher inverter gain of 114.28 V/V was achieved [121].

The two TFTs in the inverter can be stacked up or laid side-by-side, as shown in Figure 5.39 [116]. The inverted, staggered, tri-layer a-IGZO TFT was fabricated after the poly-Si was completed. The former was also treated with a dehydration step. The VTC curves of both inverters are similar. The stacked structure occupied 16% less footage than the side-by-side structure did. On the other hand, the stacked structure had a larger parasitic capacitance, which lowered the operation frequency.

The LTPO process is complicated. For example, the hydrogenation step is necessary for preparing the poly-Si TFT, e.g., to reduce the I_{off} and to increase the μ_{eff} [122]. However, hydrogen can be unintentionally released during oxide TFT deposition or post-deposition, which causes the negative shift of V_{th} or the high I_{off} [122–125]. Several methods can be applied to

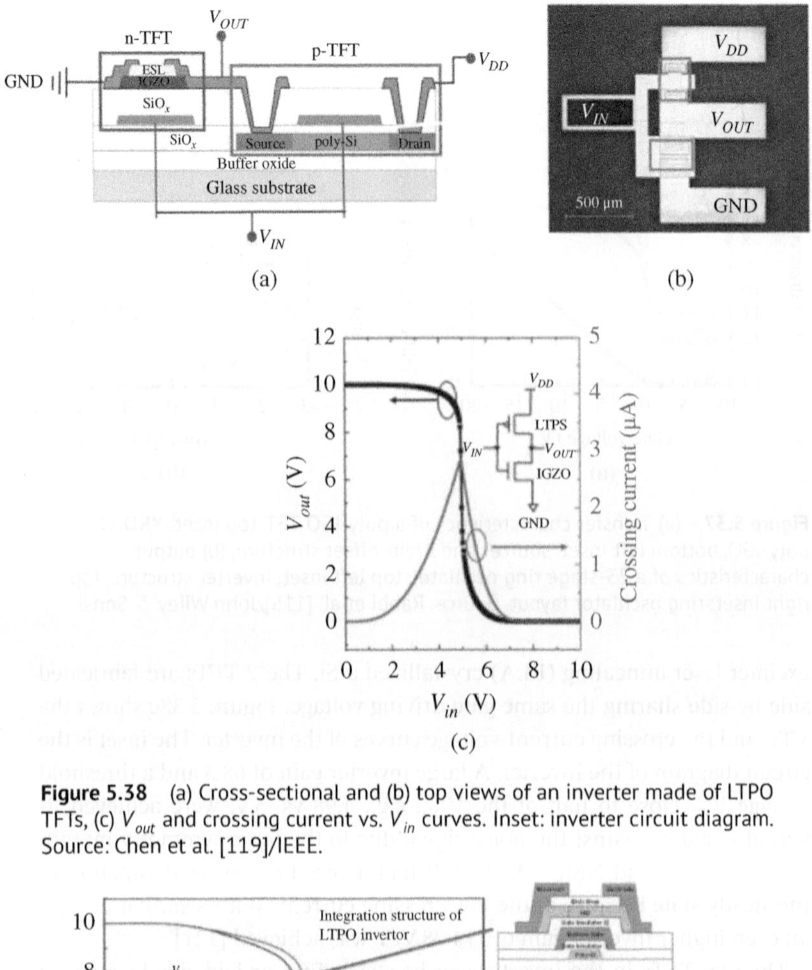

Figure 5.38 (a) Cross-sectional and (b) top views of an inverter made of LTPO TFTs, (c) V_{out} and crossing current vs. V_{in} curves. Inset: inverter circuit diagram. Source: Chen et al. [119]/IEEE.

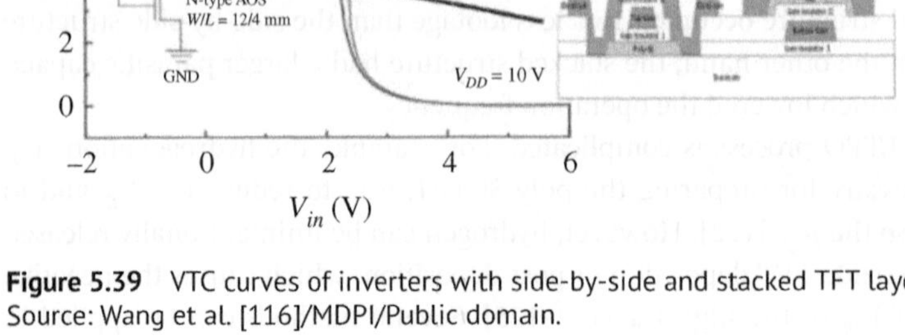

Figure 5.39 VTC curves of inverters with side-by-side and stacked TFT layouts. Source: Wang et al. [116]/MDPI/Public domain.

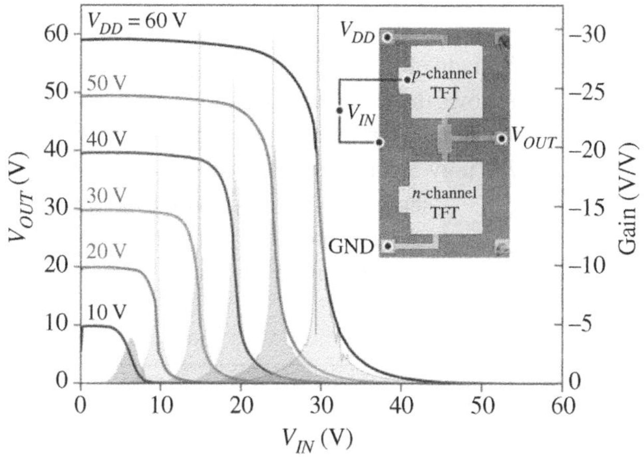

Figure 5.40 VTC and signal gain curves of an inverter composed of an n-channel oxide TFT and a p-channel organic TFT. Inset: side-by-side layout of TFTs. Source: Isakov et al. [120]/AIP Publishing.

solve the above problem. For example, low-temperature annealing can remove hydrogen from the a-IGZO TFT without deteriorating the poly-Si TFT [119]. Another method of reducing the hydrogen effect is to modify the oxide semiconductor layer. For example, when the sputter deposited a-IGZO thin film was treated with CF_4 plasma followed by N_2O oxidation first, and the hydrogenation was done using a PECVD SiN_x:H capping layer on top of the a-IGZO TFT, the hydrogen-induced defects were greatly reduced and the source/drain contact resistance was reduced [126, 127].

The inverter can be made of a n-type oxide TFT and a p-type organic TFT. For example, Figure 5.40 shows the VTC and signal gain (V/V) curves of an inverter made of a bottom-gate n-channel In_2O_3/ZnO TFT and a top-gate p-channel C8-BTBT:C16IDT-BT:C60F48 TFT operated at $V_{DD} = 10$ to 60 V [120]. The inset is the layout of TFTs. Each curve has a near-ideal V_{OUT} swing with a sharp transition at $V_{IN} = V_{DD}/2$, especially at $V_{DD} \geq 20$ V. The signal gain increases with the magnitude of the V_{DD}. It is highly immune to input signal noise in the $V_{DD} \geq 20$ V region. The high gain makes the inverter suitable for circuits comprising many logic gates/stages.

To prepare the functional inverter, both the n- and p-channel TFTs should have comparable characteristics such as μ_{eff}, V_{th}, and I_{off}. For example, to achieve a high μ_{eff}, the organic TFT can be prepared with a top-gate structure with the channel properly doped [120, 128].

5.4.3 Memory Circuits

A multilevel memory circuit made of LTPO TFTs with a capacitor has been reported [129]. Figure 5.41a shows the schematic of a LTPO memory cell T_C [129]. This simple structured memory requires two complementary TFTs and one capacitor. The p-type poly-Si TFT served as the reading transistor T_{RD} and the n-type oxide TFT served as the programming transistor T_{PG}. The storage capacitor C_{RDA}, which stored the programming voltage from T_{PG}, was connected to the gate electrode of T_{RD} to form the T_C.

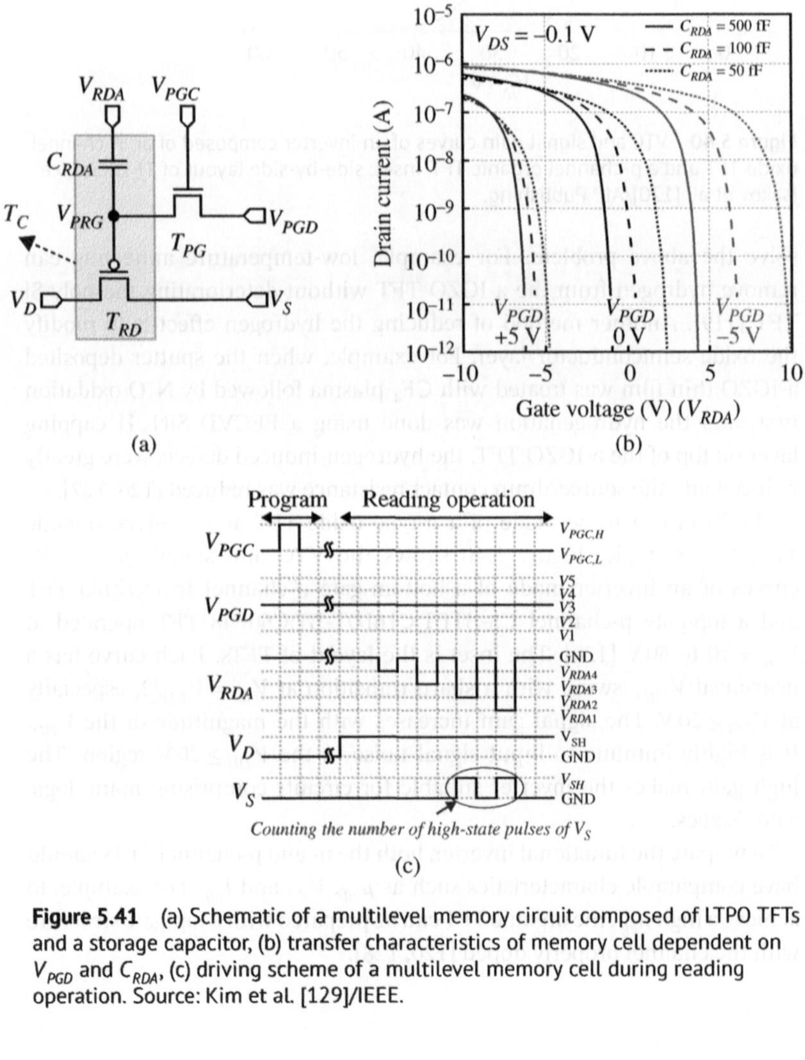

Figure 5.41 (a) Schematic of a multilevel memory circuit composed of LTPO TFTs and a storage capacitor, (b) transfer characteristics of memory cell dependent on V_{PGD} and C_{RDA}, (c) driving scheme of a multilevel memory cell during reading operation. Source: Kim et al. [129]/IEEE.

Figure 5.41b shows the transfer characteristics of the T_C concerning C_{RDA} and programming voltage V_{PDG} of T_{PG}. The threshold voltage V_{TH} of the memory cell was controlled by V_{PDG}. The V_{PGC} should be within a specific range that could completely turn on and off the T_{PG} while V_{RDA} changed. The magnitude of C_{RDA} is critical to the V_{TH} shift. Figure 5.41c shows a driving scheme of this multilevel memory cell. The magnitude of V_{RDA} determined the number of high-state pulses of V_S and therefore, the level of memory. This kind of memory device was prepared into an array for the low-power *pixel-in-memory (PIM)* circuits used in displays [130].

5.4.4 Display Drivers

For active matrix displays, such as LCDs and OLEDs, if the gate drivers are fabricated using the same TFTs as those for pixel addressing, the production cost will be lowered, the reliability will be improved, and the edge of the display will be slim. There are reports of fabricating gate drivers using a-Si:H or poly-Si TFTs [131–134]. The a-Si:H driver circuits are not suitable for high-performance displays because of the low μ_{eff} and large V_{th}. The low-temperature poly-Si TFT fabrication process can be complicated. It was reported that gate drivers could be made using n-channel amorphous oxide TFTs but suffered from anomalously high power consumption [135–140]. On the other hand, the CMOS-type drive circuit can be fabricated using LTPO TFTs [141].

The conventional gate drivers are composed of 4 TFTs and 1 capacitor, which is simple but the output signal can be unstable due to the coupling between the capacitor and the driving TFT. Circuits including more than 4 TFTs have been proposed [135, 136, 141]. Figure 5.42a shows a gate driver scheme designed for the AMOLED [141]. The inset is the cross-sectional view of the LTPS TFT and the a-IGZO TFT. Each stage consisted of four LTPS TFTs (T1, T2, T3, and T5) and two a-IGZO TFTs (T4 and T6) without any capacitor. Each stage was connected to two clock signals. CLKs and $S(n)$ varied from VSS to VL. Two AC bus lines (CLK1 and CLK2) and three DC bus lines (VSS, VBH, and VBL) were utilized in the circuit. There were three driving periods: charge, hold, and discharge. The power consumption of the circuit was dependent on the V_{TH} of the a-IGZO TFTs of the DC bus lines (VGL and VGH) and the AC bus lines (CLK1 and CLK2), as shown in Figure 5.42b. The power consumption was reduced with the decrease of the V_{TH} of the a-IGZO TFT. For example, the power was less than 100 mW at $V_{TH} = -6$ V and 2.12 mW at $V_{TH} = 0.5$ V. They are lower than that of the circuit made only of a-IGZO TFTs but higher than that of the circuit made of poly-Si TFTs. This kind of gate driver circuit can be applied to a 240 Hz 8k

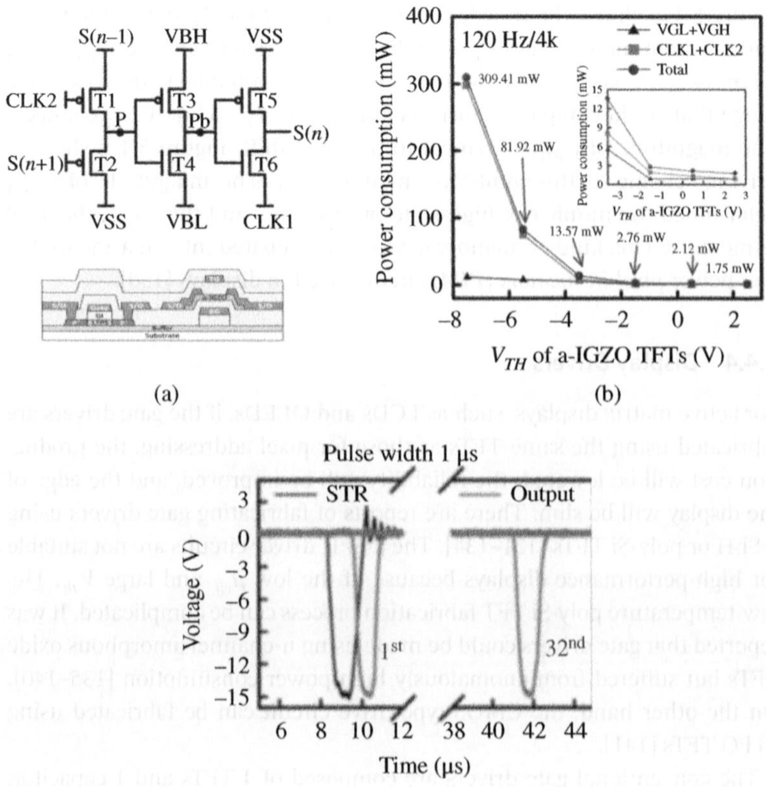

Figure 5.42 (a) A gate driver scheme composed of four LTPS TFTs and two a-IGZO TFTs, Inlet: LTPO TFT structure, (b) power consumption vs. V_{TH} of a-IGZO TFTs, (c) measured output at 1 μs pulse width. Source: Kim and Jang [141]/John Wiley & Sons.

(7,680 × 4,320) display, corresponding to one line time of 1 μs. Figure 5.42c shows the rising and falling times ~230 and ~300 ns at a pulse width of 1 μs.

5.4.5 Logic Circuits

Oxide TFT-based inverters and ring oscillators have been discussed previously in Section 5.24. The radio frequency identification (RFID) circuit is a logic application that involves not only ring oscillators but also transponder tags and rectifiers. Figure 5.43 shows an example of a RFID circuit made of bottom-gate a-IGZO TFTs on a 25 μm thick PEN substrate [142]. The maximum process temperature was 150 °C. Figure 5.43a shows the stage delay

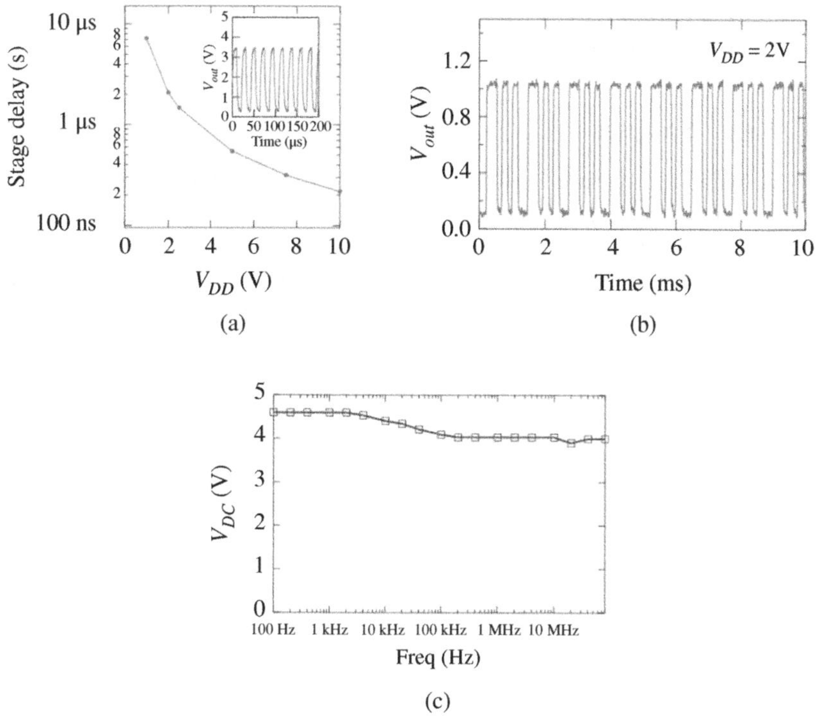

Figure 5.43 (a) Stage delay of a 19-stage ring oscillator vs. V_{DD}. Inset: time-dependent V_{OUT} pattern at $V_{DD} = 5$ V, (b) V_{OUT} pattern of an 8-bit RFID transponder circuit at $V_{DD} = 2$ V, (c) output voltage V_{DC} vs. input frequency of a rectifier at 5 V. Source: Tripathi et al. [142]/AIP Publishing.

(τ_d) of an oxide TFT-based 19-stage ring oscillator with the change of the supply voltage V_{DD}. The inset is the V_{out} output pattern at $V_{DD} = 5$ V. With a supply voltage of 2 V, the oscillation frequency reached 11 kHz corresponding to a stage delay of ~2.45 µs. As the supply voltage increased to 10 V, the oscillation frequency was as high as 109 kHz and the stage delay was reduced to ~0.24 µs. The real logic functionality is demonstrated with the functional code generators. Figure 5.43b shows the output pattern of an 8-bit RFID transponder circuit operated at $V_{DD} = 2$ V. It consisted of about 300 a-IGZO TFTs with the 5 µm design rule. The data transfer rate was 6.4 kb/s at the programmed bit pattern of 01010011.

In a passive RFID tag, the circuit is powered via wireless energy transfer from a base station sending electromagnetic radiation. The AC coupled in an antenna is changed to DC through a rectifier. Figure 5.43c shows the output voltage V_{DC} of a rectifier, which is made of diode-connected a-IGZO TFTs

($W/L = 12{,}600/5\,\mu m$), at various input frequencies. The rectified 4 V is sufficient to drive the transponder chip at the carrier frequency of 13.56 MHz.

5.4.6 Synapses

A synapse structure transmits an electrical or chemical signal from a neuron (or nerve cell) to another neuron, as shown in Figure 5.44a [144]. A neuromorphic computing system is composed of synapse-function-like devices that are operated under very low power consumption conditions and, therefore, can be applied to IoT, Machine Learning, human interface,

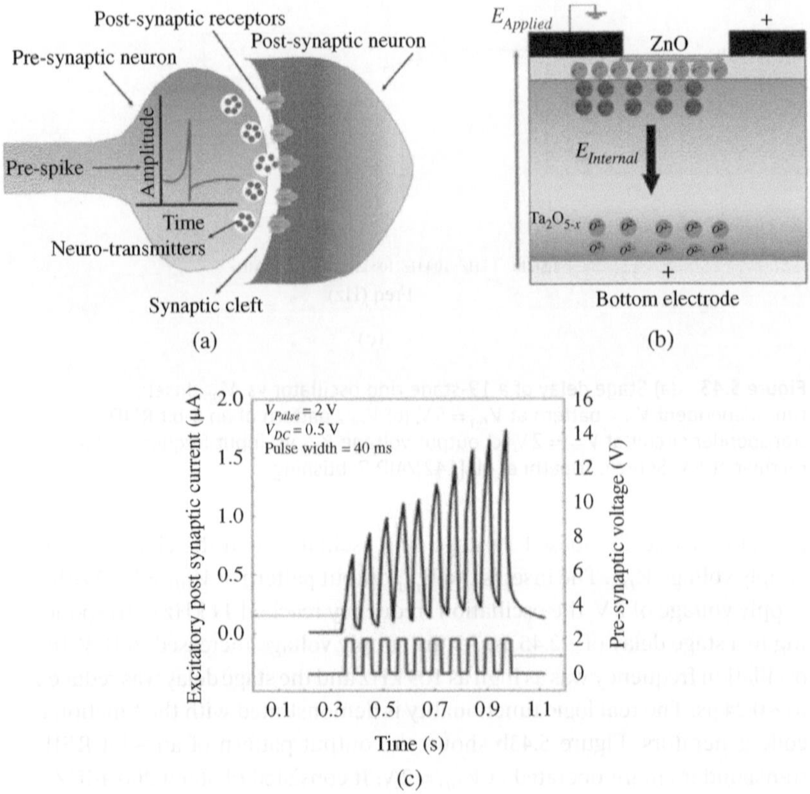

Figure 5.44 (a) A neural synapse structure [142], (b) a synapse-like function based on charge accumulation in an a-ZnO TFT, (c) EPSC response of a ZnO TFT ($W/L = 300/10\,\mu m$) to 10 presynaptic pulses of $V_g = 2$ V. Source: Pillai and DeSouza [143]/American Chemical Society.

and many other new areas. It can imitate complex information processing tasks such as image recognition, and natural language processing. Synaptic learning is judged by short-term memory (STM), e.g., from a few seconds up to tens of minutes, to long-term memory (LTM), e.g., from a few hours or a lifetime, transition. Long-term memory, i.e., long-term potentiation (LTP), could be achieved and facilitated by the repetitive electric stimulus. Plasticity is an important property of a synapse [145]. It can be expressed as synaptic potentiation, i.e., strengthening with time, and depression, i.e., weakening with time.

Metal oxide TFTs made of various types of materials or structures can function as synaptic neuron transmission devices. Charge carriers, such as oxygen vacancies or ions, can be generated, sustained, and transferred in the channel layer, as shown in Figure 5.44b [143]. Under the $+V_g$ bias condition, a positively charged oxygen-vacancies-rich region is formed in T_2O_5 adjacent to the a-ZnO channel interface. At the same time, the electron-rich interface is formed in the a-ZnO channel. When the $-V_g$ is applied, the electron accumulation in the a-ZnO disappears.

When a presynaptic spike is applied on the gate of the TFT as an external stimulus, a corresponding excitatory postsynaptic current (EPSC) peak, which is the source/drain output current, is detected. This is an analogy to biological synapse behavior. The height and shape of the EPSC peak vary with the magnitude of the $+V_g$ pulse as well as the material properties and thicknesses of the gate dielectric and channel layers. The decay data can be fitted with the following Kohlrausch equation [143].

$$G = (G_0 - G_\infty)e^{-\left(\frac{t}{\tau}\right)^\beta} + G_\infty \tag{5.3}$$

where t is the time after the end of the pre-synaptic spike, τ is the retention time, G_0 is the channel conductance at $t = 0$, G_∞ is the resting current, and β is the stretch index, typically between 0 and 1.

Figure 5.44c shows the EPSC response of a ZnO TFT ($W/L = 300/10\,\mu m$) to 10 presynaptic pulses of $V_g = 2\,V$ (width 40 ms) and $V_{DS} = 0.5\,V$ [143]. The bottom part of the figure is the gate pulse pattern applied to the TFT. A large synaptic response (~2 μA) for a narrow spike width of 40 ms has been detected. Separately, it was observed that the EPSC value and the retention time could be strongly dependent on the presynaptic pulse width, frequency, and magnitude. The synapse function can be enhanced by embedding nano Al particles in the a-IGZO channel of the TFT [146]. This kind of device showed long-term stability after being tested repeatedly with potentiating and depressing (P/D) pulses.

5.5 Oxide TFT Flexible Electronics

There are many studies on oxide TFT applications in flexible electronics for displays, circuits, sensors, etc. [147, 148]. The substrate is critical to flexible electronic applications. It must be rigid enough to support the TFTs and attached devices but flexible to sustain mechanical bending. The stiffness of a material is judged by its Young's modulus. Figure 5.45 shows the Young's modulus spectrum of materials [149].

Plastic substrates, such as polyimide (PI), PEN, polycarbonate, polyethylene terephthalate (PET), Kapton, and polyurethane, are often used in flexible electronics. Their Young's moduli are in the range of GPa. Materials, such as papers and metal foils, are flexible but are not commonly used as the substrates for electronics due to complicated reliability and processing issues. Except for the PI, most plastic substrates have glass transition temperatures (T_g's) lower than 200 °C, which limits the fabrication condition for TFTs. On the other hand, Young's moduli of metal oxides used in oxide TFTs, such as Al_2O_3 and β-Ga_2O_3, are in the range of 200–300 GPa [150, 151]. Therefore, the stress difference between the plastic substrate and the oxide TFT is very large, which needs to be minimized. The general approach of solving this kind of problem is to apply a buffer layer between the TFT and the substrate or a certain passivation layer on top of the TFT [152, 153]. The layout of the TFT can also be optimized to minimize the stress induced to the channel region during the bending test. The detailed discussion on the fabrication processes of oxide TFTs on flexible substrates is included in Chapter 4. In the following sections, the mechanical bending effects on the performance of oxide TFTs fabricated on various flexible substrates are reviewed and discussed.

5.5.1 Bending Effect on Oxide TFT Performance

The magnitude and direction of bending affect the TFT performance, as shown in Figure 5.46 [98]. Figure 5.46a is the cross-sectional view

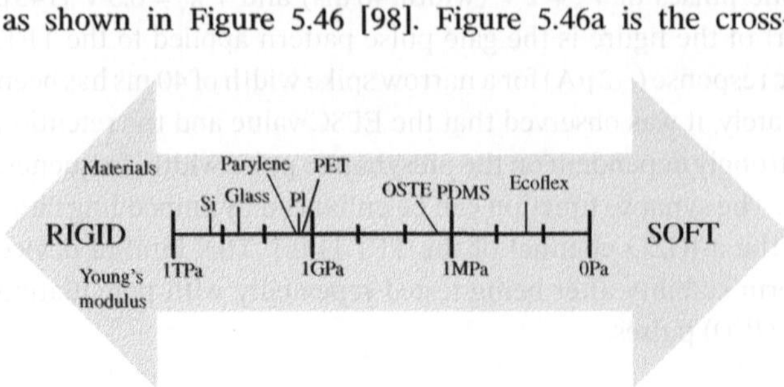

Figure 5.45 Young's modulus spectrum of materials. Source: Fallahi et al. [149]/MDPI/CC-BY-4.0.

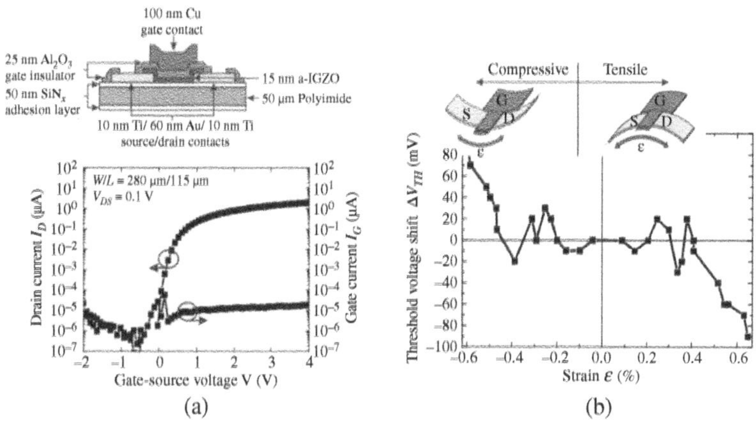

Figure 5.46 (a) Structure and transfer characteristics of an a-IGZO TFT on a PI substrate, (b) V_{th} shift vs. applied compressive and tensile strains. Source: Petti et al. [98]/with permission of IEEE.

of a top-gate TFT on a flexible PI substrate and transfer characteristics including the gate leakage current (I_g) [98]. The a-IGZO (15 nm), Al_2O_3 gate dielectric (25 nm), Ti/Au/Ti (10/60/10 nm) source and drain electrodes, and Cu (100 nm) gate electrode layers were all sputter deposited. The high-conductivity Cu gate line was connected to the IC drivers. The same SiN_x layers were deposited on the opposite sides of the PI substrate to serve as the diffusion barriers and to balance the stress on the substrate. Figure 5.46b shows the V_{th} shift with respect to the strain in the compressive and tensile stress conditions. The applied bending radii range from infinity (flat substrate) to 5.5 mm, which corresponds to the tensile strain from 0.09% to 0.64% or the compressive strain from −0.04% to −0.59%. The V_{th} shift, which was compared with the V_{th} of the TFT at flat, was observed during the mechanical bending. The V_{th} shifts from tensile and compressive bending are in opposite directions. Under tensile bending, the carrier density was increased, which was accompanied by the increase of mobility and the negative shift of V_{th}. Under compressive bending, oppositive changes in device parameters were observed [154]. When the bending radius was less than 4 mm, cracks in the TFT channel were observed, which permanently damaged the device.

The curvature of mechanical bending can influence the TFT device characteristics. However, the effect can be minimized when a proper buffer layer is included between the oxide TFT and the substrate. For example, Figure 5.47a shows the cross-section view of an inverted, staggered bi-layer a-IGZO TFT made on an organic SU-8 buffer layer coated above a 50-μm-thick PEN layer glued with a pressure-sensitive adhesive (PSA) to an alkali-free glass [152]. The maximum process temperature was 160 °C.

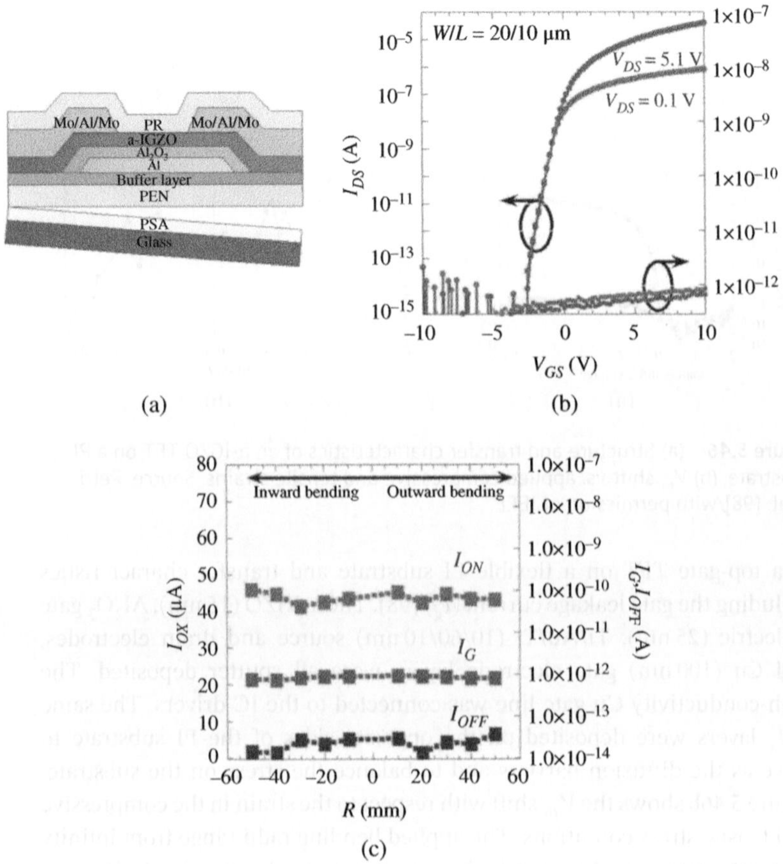

Figure 5.47 (a) Cross-sectional view of a bi-layer IGZO TFT on SU-8/PEN lifted off from PSA/glass, (b) transfer characteristics of the TFT after separation from glass, (c) I_{on}, I_{off}, and I_g of TFT vs. radius of curvature R of inward and outward bending. Source: Xu et al. [152]/IOP Publishing.

The Al_2O_3 gate dielectric layer was formed from the anodization of the aluminum (Al) gate electrode. Figure 5.47b shows the transfer characteristics of the TFT after lifting from the glass substrate on a flat surface. It has a $\mu_{sat} = 11.2\,cm^2/V\,s$, $V_{th} = 0.5\,V$, $SS = 0.27\,V/decade$, $I_{off} = 10^{-11}\,A$, and I_{on}/I_{off} ratio of 10^9. Figure 5.47c shows the I_{on}, I_{off}, and I_g of the TFT are not affected by the bending curvature R of the substrate in the inward and outward directions due to the existence of the SU-8 buffer layer.

Papers containing organic, e.g., cellulose fibers, and inorganic, e.g., pigments and fillers, components have been used as flexible substrates

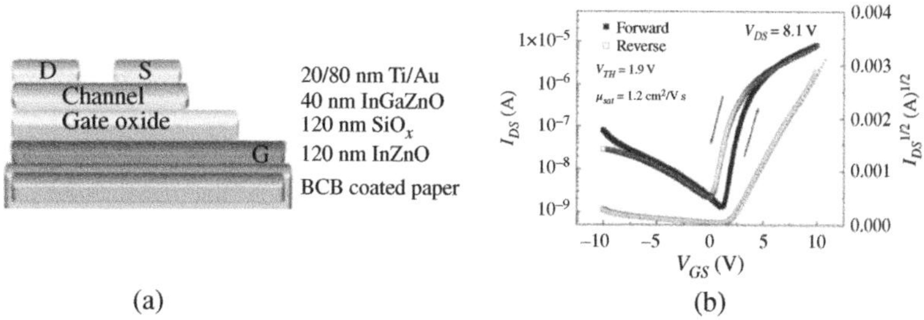

Figure 5.48 (a) An inverted, staggered a-IGZO TFT on a BCB-coated paper, (b) transfer characteristics. Source: Lim et al. [155]/AIP Publishing.

for oxide TFTs [155]. Figure 5.48 shows (a) a cross-sectional view of an inverted, staggered IGZO TFT on top of a B-staged bisbenzocyclobutene (BCB) polymer film-coated paper and (b) the transfer characteristics. The BCB layer prevents the absorption of solvent and water during the lithography and etching steps. It also provides a smoother surface for the TFT fabrication. The gate, source/drain, and channel layers were sputter-deposited. The SiO_2 gate dielectric layer was deposited by PECVD. No bending test results were presented in this paper. There are some challenges in the paper substrate application. For example, the highest process temperature for the above sample was 100 °C for two hours [155]. Commonly, the low-temperature prepared TFT suffers from the high I_{off} and large SS, as shown in Figure 5.48b, because of the high defect densities in the bulk film and at the gate dielectric/channel interface. This problem needs to be solved before practical applications.

5.5.2 Bending Effect on Oxide TFT Memory Devices, Circuits, and Others

Oxide TFT memory devices on flexible substrates have been studied for reliability. For example, Figure 5.49a shows transfer characteristic curves of an a-IGZO TFT with an Al_2O_3/p(VDF-TrFE) ferroelectric gate dielectric on a flexible PI substrate at flat or being bent in the tensile direction, i.e., at a radius of 5.5 μm (ε = 0.65%) or in the compressive direction, i.e., at a radius of 5.5 μm (ε = −0.6%) [98]. The inset is an enlarged view of the curves in the logarithm scale. Under the tensile or compressive bending conditions, the curves shifted in parallel but opposite directions. The hysteresis phenomena did not disappear under the bending condition but the memory window changed with the bending direction and radius. Figure 5.49b shows the V_{th} shifts (with respect to the V_{th} of the flat sample) in the forward

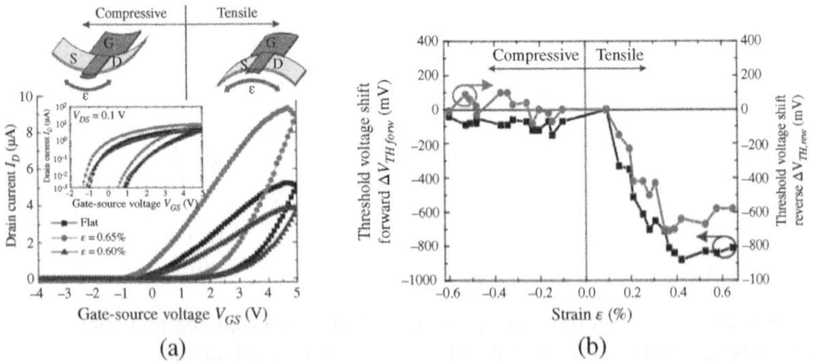

Figure 5.49 (a) Transfer characteristics of an oxide TFT memory device at flat or being bent in compressive or tensible directions, (b) threshold voltage shift of TFT memory vs. magnitude of strain. Source: Petti et al. [98]/with permission of IEEE.

($\Delta V_{TH\,forw}$) and backward ($\Delta V_{TH\,rev}$) V_g scan directions. The shift of V_{th} in the compressive bending direction was less than that in the tensile direction but the memory window of ($V_{TH,for} - V_{TH,rev}$) was not significantly affected by the bending direction. The asymmetric bending effect could be attributed to the bistable polarization of the P(VDF-TrFE) film. Since tensile and compressive strains were applied in the current flow direction, i.e., perpendicular to the direction of the dipole moments in the ferroelectric film, the tensile strain increased the polarization in the P(VDF-TrFE) layer, but the compressive strain decreased the polarization. Therefore, when the oxide TFT memory device in the flexible electronics is bent, the change in device characteristics cannot be neglected.

In a separate example, it was reported that the oxide TFT-based circuit on a flexible substrate could be operated when being bent. Figure 5.50a is the top view of a 48-stage scan driver that was made of IZO TFTs on a SiN_x (surface isolation layer SIL)/PI (20 μm thick)/buffer layer coated glass [87]. The TFT had an inverted, staggered bi-layer structure. The maximum TFT fabrication temperature was 350 °C. After the TFT ($W/L = 10/10$ μm) with the PI layer was removed from the glass substrate, the transfer characteristics were measured before bending and under bending at a curvature radius (R) varying from 10 to 40 mm. There was little change in the device characteristics except for a slight reduction of the SS and a small increase of the I_{on}/I_{off} ratio with the increase of the radius of the curvature [87]. Figure 5.50b shows the 48-stage driver circuit on top of a curved (radius of 2.5 cm) test set. There was almost no difference in the driver's characteristics before and after bending. Figure 5.50c shows output waveforms of a 10 μs-wide pulse in the 1st and 48th stages of the bent circuit measured at a clock frequency of 50 kHz with

(a) (b) (c)

Figure 5.50 (a) Structure of an IZO TFT and top view of a 48-stage scan driver, (b) photo of the circuit on a bent test set, (c) output waveforms of 1st (green line) and 48th (yellow line) stages of circuit under bending measured at 50 kHz, 5 kΩ load, 90 pf capacitance. Inset: magnified 48th stage pulse. Source: Zhang et al. [87]/IEEE.

5 kΩ load and 90 pf capacitance. The inset is the magnified view of the 48th stage pulse. The waveforms of these 2 stages are not different. Also, noises in these curves are suppressed. The circuit has been operated for 300 hours with no change of the pulse shape being observed. Therefore, when the oxide TFT-based circuit is properly prepared on a flexible substrate, it can be operated under the bending condition with satisfactory performance.

There are many other studies on oxide TFT-based flexible electronics, such as x-ray imagers [74], pH sensors [41], displays [156], inverters [157], analog circuits [158], machine learning processing engines [159], etc. In general, the flexible electronics is light in weight, easy to carry, and suitable for a wide range of applications. The flexible plastic substrate is especially useful for transparent displays. The major concern of the flexible electronics is the electric performance and reliability of the oxide TFT under the bending condition because of the mismatch of stress with the substrate and the poor characteristics of the low temperature prepared device.

5.6 Summary

Applications of oxide TFTs on various products based on the unique semiconductor material properties and device characteristics have been reviewed. While both oxide and a-Si:H TFTs can be prepared at low temperatures on large-area rigid or flexible substrates, the mobility of the oxide TFT is higher than that of the a-Si:H TFT. Therefore, the advantage of the former is the application to a much broad range of displays, circuits, sensors, etc. On the other hand, currently, oxide TFTs have lower mobilities than the poly-Si TFTs. However, they can be fabricated with simpler and lower temperature processes than the latter, which makes them suitable for the large-area substrates with a low thermal-budget requirement.

Separately, due to the large band gap energy of the oxide semiconductor material, oxide TFTs are transparent to visible light, which provides unique display and optical properties.

Examples of oxide TFT applications in LCD, OLED, μ-LED, and electrophoretic displays, solution pH, gas, temperature, pressure, fingerprint, optical, and biomolecule sensors, and imaging arrays have been presented and discussed. In addition to the conventional current-driving, oxides are available for CMOS and memory applications. Also, demonstrations of oxide TFTs in memory and logic circuits as well as synaptic and flexible electronic products have been discussed.

In summary, oxide TFTs have been proved useful in a broad range of products. For advanced circuit applications, the mobility of the n-channel oxide TFT needs to be increased. The high-performance, reliable p-channel oxide TFT must be developed for the CMOS-type circuits. Ultimately, oxide TFTs as emerging solid-state devices can be used to benefit human lives in a wide range of medical, communication, transportation, energy, environmental, entertainment, etc. areas.

References

1 Kuo, Y. (ed.) (2014). *Amorphous Silicon Thin Film Transistors*. Norwell, MA: Kluwer.

2 Kuo, Y. (ed.) (2014). *Polycrystalline Silicon Thin Film Transistors*. Norwell, MA: Kluwer.

3 Kuo, Y. and Chang, G.W. (2014). Thin film transistors as driving devices for attached devices. *ECS Transactions* 64 (10): 145–153. https://doi.org/10.1149/06410.0145ecst.

4 Kuo, Y. (1991). A new process using two masks to prepare tri-layer thin film transistors. *Journal of Electrochemical Society* 138 (2): 637–638.

5 Kuo, Y. (1995). PECVD silicon nitride as a gate dielectric for amorphous silicon thin film transistor – process and device performance. *Journal of Electrochemical Society* 142 (1): 186–190.

6 Kuo, Y. and Crowder, M. (1992). Reactive ion etching of n^+ a-Si:H: plasma damage to PECVD silicon nitride and application to thin film transistor preparation. *Journal of Electrochemical Society* 139 (2): 548–552.

7 Li, H.W., Chang, T.C., Chang, G.W. et al. (2011). Effect of lateral body terminal on silicon-oxide-nitride-oxide-silicon thin-film transistors. *IEEE Electron Device Letters* 32 (10): 1394–1396.

8 Chang, G.W., Chang, T.-C., Jhu, J.-C. et al. (2012). Suppress temperature instability of InGaZnO thin film transistors by N_2O plasma treatment, including thermal-induced hole trapping phenomenon under gate bias stress. *Applied Physics Letters* 100 (18): 182103-1.

9 Servati, P., Prakash, S.A., and Nathan, C.P. (2002). Amorphous silicon driver circuits for organic light-emitting diode displays. *Journal of Vacuum Science and Technology A* 20 (4): 1374–1378.

10 Schmidt-Mende, L., Fechtenkötter, A., Müllen, K. et al. (2001). Self-organized discotic liquid crystals for high-efficiency organic photovoltaics. *Science* 293: 1119–1122.

11 Chiang, W.J., Chen, C.-Y., Lin, C.-J. et al. (2007). Late-news poster: silicon nanocrystals photo sensor integrated on low-temperature polycrystalline-silicon panels. *SID Symposium Digest of Technical Papers* 38 (1): 294.

12 Mariucci, L., Fortunato, G., Pecora, A. et al. (1992). Hydrogenated amorphous silicon technology for chemically sensitive thin-film transistors. *Sensors and Actuators B* 6: 29–33.

13 Han, K.R., Kim, C.S., Kang, K.T. et al. (2002). Study on sensing properties of tin oxide CO gas sensor with low power consumption. *Sensors and Actuators B* 81: 182–186.

14 Fields, L.L., Zheng, J.P., Cheng, Y. et al. (2006). Room-temperature low-power hydrogen sensor based on a single tin dioxide nanobelt. *Applied Physics Letters* 88: 263102.

15 Osada, T., Akimoto, K., Sato, T. et al. (2010). Development of liquid crystal display panel integrated with drivers using amorphous In-Ga-Zn-oxide thin film transistors. *Japanese Journal of Applied Physics* 49: 03CC02.

16 Kim, B.S., Jeong, Y.T., Lee, D. et al. (2013). Solution-processed zinc-indium-tin oxide thin-film transistors for flat-panel displays. *Applied Physics Letters* 103: 072110.

17 Lee, J.-H., Kim, D.-H., Yang, D.-J. et al. (2008). World's largest (15-inch) XGA AMLCD panel using IGZO oxide TFT. *SID Symposium Digest of Technical Papers* 39 (1): 625–628.

18 Lu, H.-H., Ting, H.-C., Shih, T.-H. et al. (2008). 32-Inch LCD panel using amorphous indium-gallium-zinc-oxide TFTs. *SID Symposium Digest of Technical Papers* 41 (1): 1136–1138.

19 Hirao, T., Furuta, M., Hiramatsu, T. et al. (2008). Bottom-gate zinc oxide thin-film transistors (ZnO TFTs) for AM-LCDs. *IEEE Transactions of Electronic Devices* 55 (11): 3136–3141.

20 Morosawa, N., Nishiyama, M., Ohshima, Y.A. et al. (2014). High-mobility self-aligned top-gate oxide TFT for high-resolution AM-OLED. *Journal of Society for Information Display* 21 (11): 467–473.

21 Zhu, H., Shin, E.-S., Liu, A. et al. (2019). Printable semiconductors for backplane TFTs of flexible OLED displays. *Advanced Functional Materials* 30 (20): 1904588.

22 Verschueren, L., Ameys, M., De Roose, F. et al. (2018). Current variation reduction enabled by an external VT-compensation scheme for AMOLED displays using a 3T2C-pixel circuit with dual-gate TFTs. *SID Symposium Digest of Technical Papers* 49 (1): 437–440.

23 Cheng, S.-S., Hung, S.-C., Liu, C.-Y. et al. (2013). Fast transient V_{th} compensation circuit for AMOLED displays with P-type GOA driving. *SID Symposium Digest of Technical Papers* 44 (1): 939–942.

24 Fan, Z.Y., Lin, J.Y., Jiang, H.X. et al. (2008). III-nitride micro-emitter arrays: development and applications. *Journal of Physics D* 41 (9): 094001.

25 Jiang, H.X. and Lin, J.Y. (2013). Nitride micro-LEDs and beyond – a decade progress review. *Optics Express* 21 (3): A475–A484.

26 Zhou, X., Tian, P., Sher, C.-W. et al. (2020). Growth, transfer printing and colour conversion techniques towards full-colour micro-LED display. *Progress of Quantum Electronics* 71: 100263.

27 Fan, J., Lee, C.-Y., Chen, S.-J. et al. (2019). A RGB chip full color active matrix micro-LEDs transparent display with IGZO TFT backplane. *International Conference of Display Technology (ICDT 2019)* 50 (S1): 326–328.

28 Shin, W.-S., Ahn, H.-A., Na, J.-S. et al. (2017). A driving method of pixel circuit using a-IGZO TFT for suppression of threshold voltage shift in AMLED displays. *IEEE Electron Device Letters* 38 (6): 760–762.

29 Lin, Y.-Z., Liu, C., Zhang, J.-H. et al. (2021). Active-matrix micro-LED display driven by metal oxide TFTs using digital PWM method. *IEEE Transactions of Electron Devices* 68 (11): 5656–5661.

30 Um, J.G., Jeong, D.Y., Jung, Y. et al. (2019). Active-matrix GaN μ-LED display using oxide thin-film transistor backplane and flip chip LED bonding. *Advanced Electronic Materials* 5: 1800617. (8 p).

31 Kazlas, P.T. and McCreary, M.D. (2002). Paper-like microencapsulated electrophoretic materials and displays. *MRS Bulletin* 27 (11): 894.

32 Ito, M., Kon, M., Miyazaki, C. et al. (2008). Amorphous oxide TFT and their applications in electrophoretic displays. *Physica Status Solidi (A)* 205 (8): 1885–1894.

33 Kuo, Y. (2004). Chapter 11, Non-LCD applications of a-Si:H TFTs. In: *Amorphous Silicon Thin Film Transistors* (ed. Y. Kuo), 501–505. Norwell: Kluwer.

34 Mariucci, L., Pecora, A., Puglia, C.C. et al. (1990). Pd-gate a-Si:H thin-film transistors as hydrogen sensors. *Japanese Journal of Applied Physics* 29 (12): L2357.

35 Le Bihan, F., Carvou, E., Fortin, B. et al. (2001). Realization of polycrystalline silicon magnetic sensors. *Sensors and Actuators A* 88: 133–138.

36 Bergveld, P. (2003). Thirty years of ISFETOLOGY: what happened in the past 30 years and what may happen in the next 30 years. *Sensors and Actuators B: Chemical* 88 (1): 1–20.

37 Chi, L.-L., Chou, J.-C., Chung, W.-Y. et al. (2000). Study on extended gate field effect transistor with tin oxide sensing membrane. *Materials Chemistry and Physics* 63: 19–23.

38 Spijkman, M.-J., Myny, K., Smits, E.C.P. et al. (2011). Dual-gate thin-film transistors, integrated circuits and sensors. *Advanced Materials* 23: 3231–3242.

39 Takechi, K., Iwamatsu, S., Yahagi, T. et al. (2014). Sensitivity evaluation on amorphous $InGaZnO_4$ thin-film transistor pH sensors having various capacitances of ion-sensitive and bottom-gate insulators. *ECS Journal of Solid State Science and Technology* 3 (9): Q3076–Q3080.

40 Porwal, A. and Sahu, C. (2022). Fabrication and pH sensing characteristics measurement of back gate ZnO thin film planar FET. *Silicon* 14: 11687–11698.

41 Hyun, T. and Cho, W. (2023). Fully transparent and highly sensitive pH sensor based on an a-IGZO thin-film transistor with coplanar dual-gate on flexible polyimide substrates. *Chemosensors* 11 (1): 46. (12 p).

42 Lower, S. (2024). Chapter 16.4. The nernst equation. electrochemical chemistry. In: *Chem1 Virtual Textbook*. LibreTexts https://chem.libretexts.org/Bookshelves/General_Chemistry/Chem1_(Lower)/16%3A_Electrochemistry/16.04%3A_The_Nernst_Equation (accessed 31 July 2024).

43 Kim, K.S., Ahn, C.H., Jung, S.H. et al. (2018). Toward adequate operation of amorphous oxide thin-film transistors for low-concentration gas detection. *ACS Applied Material Interfaces* 10: 10185–11019.

44 Andringa, A.-M., Piliego, C., Katsouras, I. et al. (2014). NO_2 Detection and real-time sensing with field-effect transistors. *Chemistry of Materials* 26: 773–785.

45 Liu, P.-C., Su, C.-J., Li, P.-W. et al. (2023). Monolithically-integrated polysilicon/oxide semiconductor hybrid thin-film transistors for advanced sensing. *IEEE Journal of Electron Devices Society* 11: 490–496.

46 Chen, T.-C., Kuo, Y., Chang, T.-C. et al. (2017). Mechanism of *a*-IGZO TFT device deterioration – illumination light wavelength and substrate temperature effects. *IOP Journal of Physics D* 50: 42LT02. (6 pp).

47 Park, M.-J., Jeong, H.-S., Joo, H.-J. et al. (2019). Improvement of NO_2 gas-sensing properties in InGaZnO thin-film transistors by a pre-biasing measurement method. *Semiconductor Science and Technology* 34: 065010. (10 p).

48 Shaji, M., Saji, K.J., Jayaraj, M.K. et al. (2022). Low temperature operated ZTO thin film transistor based gas sensor for selective detection of H_2S. *Material Science in Semiconductor Processing* 150: 106927. (7 p).

49 Kim, K.S., Ahn, C.H., Kang, W.J. et al. (2017). An all oxide-based imperceptible thin-film transistor with humidity sensing properties. *Materials* 10: 530. (10 p).

50 Chen, Y.-C., Chang, T.-C., Li, H.-W. et al. (2012). High-stability oxygen sensor based on amorphous zinc tin oxide thin film transistor. *Applied Physics Letters* 100: 262908. (3 p).

51 Ghosh, S. and Rajan, L. (2021). Zinc oxide thin-film transistor with catalytic electrodes for hydrogen sensing at room temperature. *IEEE Transactions of Nanotechnology* 20: 303–310.

52 Sturm, J.C., Wilson, W., and Iodice, M. (1998). Thermal effects and scaling in organic light-emitting flat-panel displays. *IEEE Journal of Selected Topics in Quantum Electronics* 4 (1): 75–82.

53 Chen, C., Abe, K., and Kumomi, H. (2009). Density of states of a-InGaZnO from temperature-dependent field-effect studies. *IEEE Transactions on Electron Devices* 56 (6): 1177–1183.

54 Jeong, H., Kong, C.S., Chang, S.W. et al. (2013). Temperature sensor made of amorphous indium–gallium–zinc oxide TFTs. *IEEE Electron Device Letters* 34 (12): 1569–1571.

55 Park, J., Jeong, C.-Y., Cho, I.-T. et al. (2012). Fabrication of amorphous InGaZnO thin-film transistor-driven flexible thermal and pressure sensors. *Semiconductor Science and Technology* 27: 105019 (6pp).

56 Vishniakou, S., Lewis, B., Niu, X. et al. (2013). Tactile feedback display with spatial and temporal resolutions. *Science Reports* 3: 2521.

57 Vishniakou, S., Chen, R., Ro, Y.G. et al. (2018). Improved performance of zinc oxide thin film transistor pressure sensors and a demonstration of a commercial chip compatibility with the new force sensing technology. *Advanced Material Technology* 3: 1700279. (9 p).

58 Geng, D., Han, S., Seo, H. et al. (2017). Piezoelectric pressure sensing device using top-gate effect of dual-gate a-IGZO TFT. *IEEE Sensors Journal* 17 (3): 585–586.

59 Nguyen, A.H.-T., Nguyen, M.-C., Cho, S. et al. (2020). Double-gate thin film transistor with suspended-gate applicable to tactile force sensor. *Nano Convergence* 7: 31. (6 p).

60 Leland, D.S., Ginocchio, C.C. et al. (2007). Role of cell culture for virus detection in the age of technology. *Clinical Microbiology Review* 20 (1): 49–78.

61 Saylan, Y., Erdem, Ö., Ünal, S. et al. (2019). An alternative medical diagnosis method: biosensors for virus detection. *Biosensors* 9: 65. (22 p).

62 Syu, Y.-C., Hsu, W.-E., Lin, C.-T. et al. (2018). Review – field-effect transistor biosensing: devices and clinical applications. *ECS Journal of Solid State Science and Technology* 7 (7): Q3196–Q3207.

63 Grayson, A.C.R., Shawgo, R.S., Johnson, A.M. et al. (2004). A BioMEMS review: MEMS technology for physiologically integrated devices. *Proceedings of IEEE* 92 (1): 6–21.

64 Davis, F., Higson, S.P.J. et al. (2005). Structured thin films as functional components within biosensors. *Biosensors and Bioelectronics* 21: 1–20.

65 Guo, D., Zhuo, M., Zhang, X. et al. (2013). Indium-tin-oxide thin film transistor biosensors for label-free detection of avian influenza virus H5N1. *Analytica Chimica Acta* 773: 83–88.

66 Nathan, A., Lee, S., Jeon, S. et al. (2014). Amorphous oxide semiconductor TFTs for displays and imaging. *Journal of Display Technology* 10 (1): 917–927.

67 Lu, H., Zhou, X., Liang, T. et al. (2017). Oxide thin-film transistors with IMO and IGZO stacked active layers for UV detection. *IEEE Journal of the Electron Devices Society* 5 (6): 504–508.

68 Takechi, K., Nakata, M., Eguchi, T. et al. (2009). Comparison of ultraviolet photo-field effects between hydrogenated amorphous silicon and amorphous $InGaZnO_4$ thin-film transistors. *Japanese Journal of Applied Physics* 48: 010203. (3 p).

69 Görrn, P., Lehnhardt, M., Riedl, T. et al. (2007). The influence of visible light on transparent zinc tin oxide thin film transistors. *Applied Physics Letters* 91: 193504.

70 Chen, T.-C., Chang, T.-C., Hsieh, T.-Y. et al. (2010). Light-induced instability of an InGaZnO thin film transistor with and without passivation layer formed by plasma-enhanced-chemical-vapor-deposition. *Applied Physics Letters* 97: 192103.

71 Jeon, S., Song, I., Lee, S. et al. (2014). Origin of high photoconductive gain in fully transparent heterojunction nanocrystalline oxide image sensors and interconnects. *Advanced Materials* 26 (41): 7102–7109.

72 Wang, Y., Chen, C., Zou, T. et al. (2020). Spin-on-patterning of Sn-Pb perovskite photodiodes on IGZO transistor arrays for fast active-matrix near-infrared imaging. *Advanced Materials Technology* 5: 1900752.

73 Lu, J.P., Van Schuylenbergh, K., Ho, J. et al. (2002). Flat panel imagers with pixel level amplifiers based on polycrystalline silicon thin-film transistor technology. *Applied Physics Letters* 80: 4656–4658.

74 Lujan, R.A. and Street, R.A. (2012). Flexible X-ray detector array fabricated with oxide thin-film transistors. *IEEE Electron Device Letters* 33 (5): 688–690.

75 Zhao, C. and Kanicki, J. (2014). Amorphous In-Ga-Zn-O thin-film transistor active pixel sensor x-ray imager for digital breast tomosynthesis. *Medical Physics* 41 (9): 091902-1 to -14.

76 Karim, K.S., Nathan, A., Rowlands, J.A. et al. (2003). Amorphous silicon active pixel sensor readout circuit for digital imaging. *IEEE Transactions on Electron Devices* 50: 200–208.

77 Jeon, S., Park, S., Song, I. et al. (2011). Nanometer-scale oxide thin film transistor with potential for high-density image sensor applications. *Applied Material Interfaces* 3 (1): 1–6.

78 Imamura, K., Sakai, T., Takagi, T. et al. (2021). Active pixel sensor readout circuit using indium–tin–zinc-oxide thin-film transistors for image sensor applications. *Japanese Journal of Applied Physics* 60: 064002.

79 Mendis, S., Kemeny, S.E., and Fossum, E.R. (1994). CMOS active pixel image sensor. *IEEE Transactions on Electron Devices* 41: 452–453.

80 CMOS Image Sensors Market by Application and Geography - Forecast and Analysis – 2021–2025, Published: Feb 2021 Pages: 120 SKU: IRTNTR40729 Technavio. https://www.technavio.com/report/cmos-image-sensors-market-industry-analysis?source={cpc}&medium={cpc-tn-max-clicks}&gclid=CjwKCAjwquWVBhBrEiwAt1KmwnJQoa-

9BGi2q9HPGbqIwQC5cKpcV-swekZg6kcc0PJNkCMW3o2N2txoCtVA-QAvD_BwE (accessed February 2024).

81 Sakai, T., Takagi, T., Imamura, K. et al. (2021). Color-filter-free three-layer-stacked image sensor using blue/green-selective organic photoconductive films with thin-film transistor circuits on CMOS image sensors. *ACS Applied Electronic Materials* 3: 3085–3095.

82 Imamura, K., Sakai, T., Yakushiji, H. et al. (2022). Organic photoconductive film-stacked active pixel sensor pixel circuits using indium–tin–zinc-oxide thin-film transistors. *Japanese Journal of Applied Physics* 61: 075507-1 to -5.

83 Kuo, Y. (2004). Chapter 13, Poly-Si TFTs for non-LCD applications. In: *Polycrystalline Silicon Thin Film Transistors* (ed. Y. Kuo), 490–492. Norwell: Kluwer.

84 Young, N.D., Harkin, G., Bunn, R.M. et al. (1997). Novel fingerprint scanning arrays using polysilicon TFTs on glass and polymer substrates. *IEEE Electron Device Letters* 18 (1): 19–20.

85 Hashido, R., Suzuki, A., Iwata, A. et al. (2001). A capacitive fingerprint sensor with low-temperature poly-Si TFTs. *Solid-State Circuits Conferenc Digest Technical Papers*, (ISSCC 2001 IEEE International) 452: 250–251.

86 Qin, Y., Wang, H., Liu, Y. (2018). Organic-inorganic hybrid thin-film photo-detector for fingerprint recognition. *SID Symposium Digest Technical Papers* 49(1): 1604–1606.

87 Zhang, L.-R., Huang, C.-Y., Li, G.-M. et al. (2016). A low-power high-stability flexible scan driver integrated by IZO TFTs. *IEEE Transactions Electron Devices* 63 (4): 1779–1782.

88 Ou, C.-W., Dhananjay, H.,.Z.Y., Chuang, Y.-C. et al. (2008). Anomalous p-channel amorphous oxide transistors based on tin oxide and their complementary circuits. *Applied Physics Letters* 92: 122113.

89 Kuo, Y. (2013). Nanocrystals embedded high-*k* nonvolatile memories – bulk film and nanocrystal material effects. *ECS Transactions* 53 (4): 121–128.

90 Han, H., Jang, S., Kim, D. et al. (2022). Memory characteristics of thin film transistor with catalytic metal layer induced crystallized indium-gallium-zinc-oxide (IGZO). *Channel. Electronics* 11: 53. (12 pp).

91 Jeong, S., Jang, S., Han, H. et al. (2021). C-axis aligned crystalline indium-gallium-zinc oxide (CAAC-IGZO) and high-k charge trapping film for flash memory application. *Journal of Alloys and Compounds* 888: 161440. (12 pp).

92 Suresh, A., Novak, S., Wellenius, P. et al. (2009). Transparent indium gallium zinc oxide transistor based floating gate memory with platinum nanoparticles in the gate dielectric. *Applied Physics Letters* 94: 123501.

93 Miyasako, T., Senoo, M., and Tokumitsu, E. (2005). Ferroelectric-gate thin-film transistors using indium-tin-oxide channel with large charge controllability. *Applied Physics Letters* 86: 162902.

94 Prins, M.W.J., Grosse-Holz, K.-O., Müller, G. et al. (1996). A ferroelectric transparent thin-film transistor. *Applied Physics Letters* 68: 3650.

95 Prins, M.W.J., Zinnemers, S.E., Cillessen, J.F.M. et al. (1997). Depletion-type thin-film transistors with a ferroelectric insulator. *Applied Physics Letters* 70: 458–460.

96 Minh, D.H., Loi, N.V., Duc, N.H. et al. (2016). Low-temperature PZT thin-film ferroelectric memories fabricated on SiO_2/Si and glass substrates. *Journal of Science: Advanced Materials and Devices* 1: 75–79.

97 Böscke, T.S., Müller, J., Bräuhaus, D. et al. (2011). Ferroelectricity in hafnium oxide thin films. *Applied Physics Letters* 99: 102903.

98 Petti, L., Münzenrieder, N., Salvatore, G.A. et al. (2014). Influence of mechanical bending on flexible InGaZnO-based ferroelectric memory TFTs. *IEEE Transactions on Electron Devices* 61 (4): 1085–1092.

99 Kim, B.H., Byun, C.W., Yoon, S.-M. et al. (2011). Oxide-thin-film-transistor-based ferroelectric memory array. *IEEE Electron Device Letters* 32 (3): 324–326.

100 Zhao, W., Law, J., Waechter, D. et al. (1998). Digital radiology using active matrix readout of amorphous selenium: detectors with high voltage protection. *Medical Physics* 25 (4): 539–549.

101 Chow, E.M., Lu, J.P., Ho, J. et al. (2006). High voltage thin film transistors integrated with MEMS. *Sensors and Actuators A: Physical* 130–131: 297–301.

102 Gamo, H., Kanemaru, S., and Itoh, J. (1998). Fabrication of a new field emitter array with a built-in thin-film transistor on glass. *Japanese Journal of Applied Physics* 37 (1) 12B: 7134–7137.

103 Karpelson, M., Wei, G.-Y., and Wood, R.J. (2012). Driving high voltage piezoelectric actuators in microrobotic applications. *Sensors and Actuators A: Physical* 176: 78–89.

104 Martin, R.A., Da Costa, V.M., Hack, M. et al. (1993). High-voltage amorphous silicon thin-film transistors. *IEEE Transactions on Electron Devices* 40 (3): 634–644.

105 Hong, W.-C., Ku, C.-J., Li, R. et al. (2016). MgZnO high voltage thin film transistors on glass for inverters in building integrated photovoltaics. *Scientific Reports* 6: 34169.

106 Gong, Y. and Jackson, T.N. (2017). Offset drain ZnO thin-film transistors for high-voltage operation. *IEEE Electron Device Letters* 38 (8): 1047–1050.

107 Li, X., Liu, C., Kong, J. et al. (2022). Widely adjusting the breakdown voltages of kilo-voltage thin film transistors. *IEEE Electron Device Letters* 43 (2): 240–243.

108 Unagami, T. (1988). High-voltage poly-Si TFTs with multichannel structure. *IEEE Transactions on Electron Devices* 35 (12): 2363–2367.

109 Mourey, D. A., Zhao, D. A., Fok, H. H. R. et al. (2010). Thermal effects in oxide TFTs. *Proceedings of Device Research Conference.* pp. 243–244, South Bend, IN, USA (Jun. 2010).

110 Marette, A., Shea, H.R., and Briand, D. (2018). Yttrium zinc tin oxide high voltage thin film transistors. *Applied Physics Letters* 113 (13): 132101.

111 Li, X., Liu, C., Liu, C. et al. (2020). Kilo-voltage thin-film transistors for driving nanowire field emitters. *IEEE Electron Device Letters* 41 (3): 405–408.

112 Presley, R.E., Hong, D., Chiang, H.Q. et al. (2006). Transparent ring oscillator based on indium gallium oxide thin-film transistors. *Solid State Electronice* 50 (3): 500–503.

113 Ofuji, M., Abe, K., Shimizu, H. et al. (2007). Fast thin-film transistor circuits based on amorphous oxide semiconductor. *IEEE Electron Device Letters* 28 (4): 273–275.

114 Nomura, K. (2021). *Journal of Information Display* 22 (4): 211–229.

115 Rabbi, M.H., Lee, S., Sasaki, D. et al. (2022). High-performance, coplanar polycrystalline InGaO thin-film transistor for large-area, high-resolution AMOLED display. *SID 2022 Digest* 16–19.

116 Wang, Y., Zhou, Y., Xia, Z. et al. (2022). Compact integration of hydrogen-resistant a-InGaZnO and poly-Si thin-film transistors. *Micromachines* 13 (6): 839. (8 p).

117 Wang, S., Shi, R., Li, J. et al. (2020). Resilience of fluorinated indium-gallium-zinc oxide thin-film transistor against hydrogen-induced degradation. *IEEE Electron Device Letters* 41 (5): 929–732.

118 Aman, M., Takeda, Y., Ito. et al. (2021). Reliability improvement of IGZO-TFT in hybrid process with LTPS. *Journal of Society for Information Display* 29: 416–427.

119 Chen, C.D., Yang, B.-R., Liu, C. et al. (2017). Integrating poly-silicon and InGaZnO thin-film transistors for CMOS inverters. *IEEE Transactions on Electron Devices* 64 (9): 3668–3671.

120 Isakov, I., Paterson, A.F., Solomeshch, O. et al. (2016). Hybrid complementary circuits based on p-channel organic and n-channel metal oxide transistors with balanced carrier mobilities of up to $10\,cm^2/Vs$. *Applied Physics Letters* 109 (26): 263301.

121 Kim, H., Jeong, D.Y., Lee, S. et al. (2019). A high-gain inverter with low-temperature poly-Si oxide thin-film transistors. *IEEE Electron Device Letters* 40 (3): 411–414.

122 Kim, M.H., Lee, Y.-A., Kim, J. et al. (2015). Photochemical hydrogen doping induced embedded two-dimensional metallic channel formation in InGaZnO at room temperature. *ACS Nano* 9 (10): 9964–9973.

123 Toda, T., Wang, D., Jiang, J. et al. (2014). Quantitative analysis of the effect of hydrogen diffusion from silicon oxide etch-stopper layer into amorphous In-Ga-Zn-O on thin-film transistor. *IEEE Transactions on Electron Devices* 61 (11): 3762–3767.

124 Kim, J., Bang, S., Lee, S. et al. (2012). A study on H_2 plasma treatment effect on a-IGZO thin film transistor. *Journal of Materials Research* 27 (17): 2318–2325.

125 Bang, J., Matsuishi, S., and Hosono, H. (2017). Hydrogen anion and subgap states in amorphous In–Ga–Zn–O thin films for TFT applications. *Applied Physics Letters* 110: 232105.

126 Park, J., Kim, S., Kim, C. et al. (2008). High-performance amorphous gallium indium zinc oxide thin-film transistors through N_2O plasma passivation. *Applied Physics Letters* 93: 053505.

127 Knobelspies, S., Takabayashi, A., Daus, A. et al. (2018). Improvement of contact resistance in flexible a-IGZO thin-film transistors by CF_4/O_2 plasma treatment. *Solid-State Electronics* 150: 23–27.

128 Paterson, A.F., Treat, N.D., Zhang, W. et al. (2016). Small molecule/polymer blend organic transistors with hole mobility exceeding $13\,cm^2\,V^{-1}s^{-1}$. *Advanced Materials* 28: 7791–7798.

129 Kim, J., Chung, H.-J., and Lee, S.-W. (2021). Multi-level memory comprising low-temperature poly-silicon and oxide TFTs. *IEEE Electron Device Letters* 42 (1): 42–45.

130 Chang, T.-K., Lin, C.-W., Chang, S. (2022). LTPO TFT technology for AMOLEDs. *Journal of Society for Information Display* 30(3): 175–197.

131 Stewart, R.G. (1995). Circuit design for a-Si AMLCDs with integrated drivers. *SID'95* 89–92.

132 Nathan, A., Servati, P., Karim, K.S. et al. (2004). Chapter 3, Device physics, compact modeling, and circuit applications of a-Si:H TFTs. In: *Amorphous Silicon Thin Film Transistors* (ed. Y. Kuo), 79–181. Norwell: Kluwer.

133 Choi, B.-D., Lee, I., and Kwon, O.-K. Chapter 11, Poly-Si TFT drivers. In: *Polycrystalline Thin Film Transistors* (ed. Y. Kuo), 360–424. Norwell: Kluwer.

134 Song, S.-J., Kim, B.H., Jang, J. et al. (2015). Low power low-temperature poly-Si thin-film transistor shift register with DC-type output driver. *Solid-state Electronics* 111: 204–209.

135 Lin, C.-L., Du, Y.-W., Cheng, M.-H. et al. (2014). A low-power gate driver using depletion-mode a-IGZO TFTs. *SID Symposium Digest* 45 (1): 1039–1042.

136 Tu, C.-D., Tsao, S.-W., Huang, C.-H. et al. (2017). A gate driver circuit with a-IGZO TFTs for 8-inch QXGA TFT-LCD panel. *SID Symposium Digest* 48 (1): 1416–1418.

137 Kim, B., Ryoo, C.-I., Kim, S.-J. et al. (2011). New depletion-mode IGZO TFT shift register. *IEEE Electron Device Letters* 32 (2): 158–160.

138 Pi, J.-E., Ryu, M.K., Hwang, C.-S. et al. (2012). A low-power scan driver circuit for oxide TFTs. *IEEE Electron Device Letters* 33 (8): 1144–1146.

139 Yoo, S.-J., Hong, S.-J., Kang, J.-S. et al. (2012). A low-power single-clock-driven scan driver using depletion-mode a-IGZO TFTs. *IEEE Electron Device Letters* 33 (3): 402–404.

140 Wu, W.-J., Li, G.-M., Xia, X.-H. et al. (2014). Low-power bi-side scan driver integrated by IZO TFTs including a clock-controlled inverter. *Journal of Display Technology* 10 (7): 523–525.

141 Kim, J. and Jang, J. (2022). Gate driver with low-temperature poly-crystalline silicon and oxide thin-film-transistor circuits for low power consumption and narrow bezel AMOLED displays. *Journal of Society for Information Display* 30: 482–488.

142 Tripathi, A.K., Smits, E.C.P., van der Putten. et al. (2011). Low-voltage gallium-indium-zinc-oxide thin film transistors based logic circuits on thin plastic foil: Building blocks for radio frequency identification application. *Applied Physics Letters* 98: 162102.

143 Pillai, P.B. and DeSouza, M.M. (2017). Nanoionics-based three-terminal synaptic device using zinc oxide. *ACS Applied Materials Interfaces* 9: 1609–1618.

144 Oh, S., Kwon, O., Seo, H.Y. et al. (2023). Light-stimulated long-term potentiation behavior enhanced in a HfO_2/InGaZnO photonic synapse. *Applied Materials Today* 34: 101919. (8 p).

145 Hughes, J.R. (1958). Post-tetanic potentiation. *Physiology Reviews* 38 (1): 91–113.

146 Kim, J., Kim, Y., Kwon, O. et al. (2020). Modulation of synaptic plasticity mimicked in Al nanoparticle-embedded IGZO synaptic transistor. *Advanced Electronic Materials* 6: 1901072. (9 p).

147 Miyakawa, M., Tsuji, H., and Nakata, M. (2022). Highly stretchable island-structure metal oxide thin-film transistor arrays using acrylic adhesive for deformable display applications. *Journal of Society for Information Display* 30 (9): 699–705.

148 Bhalerao, S. (2021). Flexible, low-voltage, metal oxide thin film transistors (TFT) and circuits for wearables and internet of things (IoT). Dissertation, Tampere University, Faculty of Information Technology and Communication Sciences (ITC), Finland, 17th December 2021.

149 Fallahi, H., Zhang, J., Phan, H.-P. et al. (2019). Flexible microfluidics: fundamentals, recent developments, and applications. *Micromachines* 10: 830. (26 p).

150 Grosskreutz, J.C. (1969). Mechanical properties of metal oxide films. *Journal of Electrochemical Society* 116: 1232–1237.

151 Yu, M.-F., Atashbar, M.Z., and Chen, X. (2005). Mechanical and electrical characterization of β-Ga_2O_3 nanostructures for sensing applications. *IEEE Sensors Journal* 5 (1): 20–25.

152 Xu, H., Pang, J., Xu, M. et al. (2014). Fabrication of flexible amorphous indium-gallium-zinc-oxide thin-film transistors by a chemical vapor deposition-free process on polyethylene napthalate. *ECS Journal of Solid State Science and Technology* 3 (9): Q3035–Q3039.

153 Jung, T. and Jeon, S. (2022). Effects of iCVD organic passivation in oxide thin-film transistors under repetitive bending stress for electrical and mechanical stability. *Journal of Vacuum Science and Technology B* 40: 042201. (7 p).

154 Munzenrieder, N., Cherenack, K.H., and Troster, G. (2011). The effects of mechanical bending and illumination on the performance of flexible IGZO TFTs. *IEEE Transactions on Electron Devices* 58 (7): 2041–2048.

155 Lim, W., Douglas, E.A., and Norton, D.P. (2010). Low-voltage indium gallium zinc oxide thin film transistors on paper substrates. *Applied Physics Letters* 96: 053510.

156 Uesaka, S., Yamaoka, R., Sasaki, T. et al. (2015). Quad-FHD flexible CAAC-OS AMOLED display with long-life OLED device structure. *Journal of SID* 22 (12): 603–612.

157 Kim, D.I., Hwang, B.U., Park, J.S. et al. (2012). Mechanical bending of flexible complementary inverters based on organic and oxide thin film transistors. *Organic Electronics* 13 (11): 2401–2405.

158 Cantarella, G., Costa, J., Meister, T. et al. (2020). Review of recent trends in flexible metal oxide thin-film transistors for analog applications. *Flexible and Printed Electronics* 5: 033001. (24 p).

159 Ozer, E., Kufel, J., Myers, J. et al. (2020). A hardwired machine learning processing engine fabricated with submicron metal-oxide thin-film transistors on a flexible. *Nature Electronics* 3: 419–425.

Index

Oxide Thin Film Transistors, First Edition. Yue Kuo, Hideo Hosono, Michael S. Shur, and Jin Jang.

Printed and bound by CPI Group (UK) Ltd, Croydon, CR0 4YY

17/09/2026

14964309-0002